The Art of
Image Processing
with Java

Kenny A. Hunt

A K Peters, Ltd.
Natick, Massachusetts

Editorial, Sales, and Customer Service Office

A K Peters, Ltd.
5 Commonwealth Road, Suite 2C
Natick, MA 01760
www.akpeters.com

Library of Congress Cataloging-in-Publication Data

Hunt, Kenny A.
  The art of image processing with Java / Kenny A. Hunt.
    p. cm.
  Includes bibliographical references and index.
  ISBN 978-1-56881-717-0 (alk. paper)
  1. Image processing–Digital techniques. 2. Java (Computer program language)
 I. Title.

TA1637.H87 2010
621.36'702855133–dc22

2010027302

Printed in India
14 13 12 11 10   10 9 8 7 6 5 4 3 2 1

Contents

Introduction

1

1.1   What Is Digital Image Processing?

We must begin our journey by taking issue with the philosophical adage that "a picture is worth a thousand words." It is my belief that a picture cannot begin to convey the depth of human experience and wisdom embedded in the words of Shakespeare, Dostoevsky, Dante, or Moses. A picture cannot convey with due precision the mathematical underpinnings of the discoveries of Galileo or Pascal nor can a picture give expression to the philosophy of Augustine, Plato, or Edwards. Nonetheless, while pictures do not carry the precision of written language, they do contain a wealth of information and have been used throughout the centuries as an important and useful means of communication. An *image* is a picture representing visual information. A *digital image* is an image that can be stored in digital form.

Prior to the advent of computation, images were rendered on papyrus, paper, film, or canvas using ink or paint or photosensitive chemicals. These non-digital images are prone to fading and hence suffer loss of image quality due to exposure to light or temperature extremes. Also, since non-digital images are fixed in some physical medium it is not possible to precisely copy a non-digital image. Throughout the annals of art history, forgers have attempted to copy paintings of well-known masters but usually fail due to their inability to precisely duplicate either a style or an original work. Han van Meegeren is one of the best known art forgers of the 20th century. His technique so closely mimicked the style and colors of the art masters that he was able to deceive even the most expert art critics of his time. His most famous forgery, *The Disciples at Emmaus*, was created in 1936 and was purportedly created by the well-known Dutch artist Johannes Vermeer. His work was finally exposed as fraudulent, however, at least in part by a chemical analysis of the paint, which showed traces of a plastic compound that was not manufactured until the 20th century!

Digital images, however, are pictures that are stored in digital form and that are viewable on some computing system. Since digital images are stored as binary data, the digital image never fades or degrades over time and the only way to destroy a digital image is to delete or corrupt the file itself. In addition, a digital image can be transmitted across the globe in seconds and can be efficiently and precisely copied without any loss of quality.

Digital image processing is a field of study that seeks to analyze, process, or enhance a digital image to achieve some desired outcome. More formally, digital image processing can be defined as the study of techniques for transforming a digital image into another (improved) digital image or for analyzing a digital image to obtain specific information about the image.

From the cradle to the grave we are accustomed to viewing life through *digital* images. A parent's first portrait of their child is often taken before they are even born through the use of sophisticated ultrasound imaging technology. As the child grows, the parents capture developmental milestones using palm-sized digital video cameras. Portraits are sent over email to relatives and friends and short video clips are posted on the family's website. When the child breaks an arm playing soccer, the emergency-room physician orders an x-ray image and transmits it over the Internet to a specialist hundreds of miles away for immediate advice. During his lifetime the child will watch television images that have been digitally transmitted to the dish on top of his house, view weather satellite images on the Internet to determine whether or not to travel, and see images of war where smart bombs find their target by "seeing" the enemy.

Computer graphics is a closely related field but has a different goal than image processing. While the primary goal of computer graphics is the efficient generation of digital images, the input to a graphics system is generally a geometric model specifying the shape, texture, and color of all objects in the virtual scene. Image processing, by contrast, begins with a digital image as input and generates, typically, a digital image as output.

Computer vision, or machine vision, is another increasingly important relative of image processing where an input image is analyzed in order to determine its content. The primary goal of computer vision systems is the inverse of

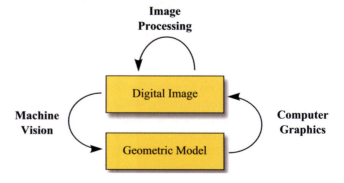

Figure 1.1. Disciplines related to image processing.

Figure 1.2. Image processing pipeline.

computer graphics: to analyze a digital image and infer meaningful information about the scene depicted by the image. Figure 1.1 illustrates the roles and relationships between each of these three disciplines where boxes represent a type of data while the connecting arrows show the typical input and output for a field of study.

A complete digital image processing system is able to service every aspect of digital image handling. Figure 1.2 shows the five typical stages in an image processing pipeline: image acquisition, image processing, image archival, image transmission, and image display. Image acquisition is the process by which digital images are obtained or generated. Image processing is the stage where a digital image is enhanced or analyzed. Image archival is concerned with how digital images are represented in memory. Image transmission is likewise concerned with data representation but places added emphasis on the robust reconstruction of potentially corrupted data due to transmission noise. Image display deals with the visual display of digital image data whether on a computer monitor, television screen, or printed page.

A visual example of the pipeline stages is given in Figure 1.3. During the image acquisition stage, an approximation of a continuous tone or analog scene is recorded. Since the captured image is an approximation, it includes some error which is introduced through sampling and quantization. During archival, a further degradation of quality may occur as the concern to conserve memory and hence conserve transmission bandwidth competes with the desire to maintain a high quality image. When the image is displayed, in this case through printing in black and white, image quality may be compromised if the output display is unable to reproduce the image with sufficient resolution or depth of color.

Construction of a complete image processing system requires specialized knowledge of how hardware architecture, the physical properties of light, the workings of the human visual system, and the structure of computational techniques affects each stage in the pipeline. Table 1.1 summarizes the most important topics of study as they correspond to each of the five primary stages in an image processing system. Of course a deep understanding of each of the listed areas of study is required to construct an efficient and effective processing module within any stage of the pipeline. Nevertheless, each stage of the processing pipeline raises unique concerns regarding memory requirements, computational efficiency, and image quality. A thorough understanding of the affects of each stage on image processing is required in order to achieve the best possible balance among memory, computation time, and image quality.

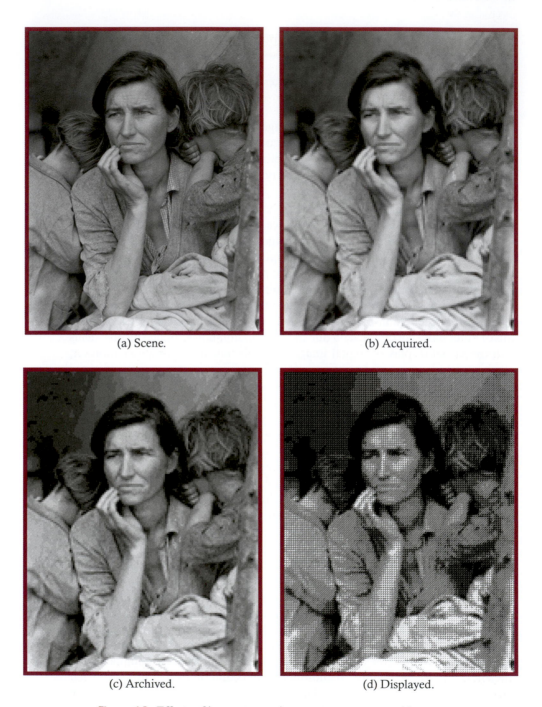

(a) Scene.   (b) Acquired.

(c) Archived.   (d) Displayed.

Figure 1.3. Effects of image processing stages on a processed image.

Processing Stage	Topic of Study
acquisition	physical properties of light human perception mathematical models of color
processing	software architecture data representation algorithm design
archival	compression techniques data representation
transmission	data representation transmission protocols
display	digital halftoning color models human perception

Table 1.1. Topics of study in image processing.

These five stages serve as a general outline for the remainder of this text. The image processing topics associated with each stage of the processing pipeline will be discussed with an emphasis on the processing stage which lies at the heart of image processing. By contrast, little coverage will be allocated to transmission issues in particular.

1.2 Why Digital Image Processing?

Digital images are used across an exceptionally wide spectrum of modern life. Ranging from digital cameras and cell phones to medical scans and web technology, digital image processing plays a central role in modern culture. This section provides examples of practical applications of image processing techniques. A general overview of these applications suffices to illustrate the importance, power, and pervasiveness of image processing techniques.

1.2.1 Medicine

Digital imaging is beginning to supplant film within the medical field. Computed tomography (CT) is a noninvasive imaging technique used to diagnose various ailments such as cancers, trauma, and musculoskeletal disorders. Magnetic resonance imaging (MRI) is a similarly noninvasive method for imaging the internal structure and function of the body. MRI scans are more amenable to diagnosing neurological and cardiovascular function than CT scans due to their greater contrast among soft tissue volumes. Figure 1.4 gives an example of both MRI and CT images where the MRI highlights contrast in the internal soft-tissue organs of a human pelvis while the CT image captures the internal skeletal structure of a human skull.

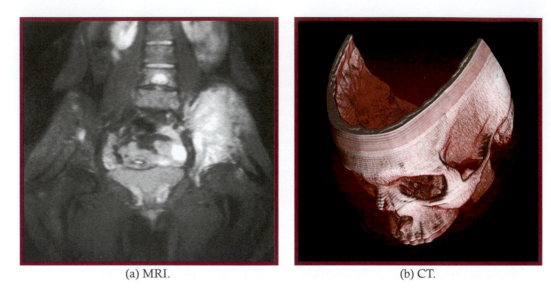

(a) MRI.   (b) CT.

Figure 1.4. Medical images.

Since errors in the acquisition, processing, archival, or display of medical images could lead to serious health risks for patients, rigorous standards have been developed to ensure that digital images for medical use are properly archived and displayed. The Digital Imaging and Communications in Medicine (DICOM) is one such standard and has become the de facto standard for image processing in the health professions.

1.2.2   Biology

Biology is a natural science that studies living organisms and how they interact with the environment. Biological research covers a vast array of specialized subdisciplines such as botany, zoology, cell biology, microbiology, and biochemistry. Each of these disciplines relies to some degree on sophisticated computing systems to acquire and analyze large amounts of image-based data. These measurements ultimately provide information required for tasks such as deciphering complex cellular processes and identifying the structure and behavior of DNA.

Since image-based measurement is becoming increasingly vital to biological research, biologists must have basic knowledge in image processing to correctly interpret and process their results. Part (a) of Figure 1.5 shows a scanning electron microscope (SEM) image of a rust mite where the length of the mite is on the order of 60 μm. Part (b) shows the structure of the eye of a fruit fly where each spherical sensor is on the order of 10 μm in diameter.

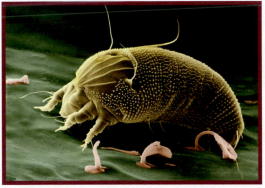 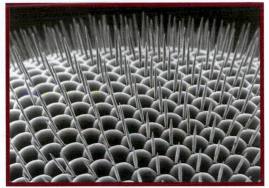

(a) A rust mite (*Aceria anthocoptes*).   (b) The eye of a fruit fly (*Drosophilidae*).

Figure 1.5. Images in biology.

1.2.3 Biometrics

The security of national, corporate, and individual assets has become a topic of great importance in the modern global economy as terrorists, con men, and white-collar criminals pose an ongoing threat to society. When a person boards an airplane, enters credit card information over the Internet, or attempts to access medical records for a hospital patient, it is desirable to verify that the person actually is who they claim to be. The field of biometrics seeks to verify the identity of individuals by measuring and analyzing biological characteristics such as fingerprints, voice patterns, gait, facial appearance, or retinal scans. In most of these techniques, with the exception of voice recognition, the biological traits are obtained by the analysis of a digital image.

Biometrics has been used for decades in law enforcement to identify criminals from fingerprint images. Highly trained experts have traditionally performed fingerprint identification manually by comparing fingerprints of criminal suspects with fingerprints obtained from a crime scene. Systems are now commonly used to match fingerprints against large databases of suspects or known criminals. Specialized hardware is used to first acquire a digital image of an individual's fingerprint. Software is then used to analyze the image and compare it with a large database of known fingerprint images. Since the process is automated, it is possible to quickly search a very large database and quickly obtain accurate verification.

The use of palm scans is proving increasingly effective in the field of biometrics. A palm scanner is used to acquire an image of the blood flow through the veins of the hand in a completely noninvasive and contact-free fashion. Since the veins form a complex three-dimensional structure within a person's palm, individuals can be identified with extremely high accuracy, and forgery is extremely difficult.

1.2.4   Environmental Science

All life depends upon a healthy environment and the environmental sciences seek to understand the forces that affect our natural world. Environmental science is a broad and interdisciplinary field that includes the study of weather patterns (meteorology), oceans (oceanography), pollution as it affects life (ecology), and the study of the earth itself (the geosciences).

Data acquisition and analysis plays a key role in each of these fields since monitoring oceans, forests, farms, rivers, and even cities is critical to proper stewardship. Computer and imaging systems play an increasingly active and central role in these tasks. Satellite imaging is used to monitor and assess all types of environmental phenomena, including the effects of wildfires, hurricanes, drought, and volcanic eruptions. Motion-sensitive cameras have been installed in remote regions to monitor wildlife population densities. In recent years, these systems have discovered many new species and have even taken photographs of animals long believed extinct.

Figure 1.6 shows two enhanced satellite images of St. Louis, Missouri. The image in Figure 1.6(a) was taken during the great flood of 1993 while the image in Figure 1.6(b) was taken the following year. Environmental scientists tracked and measured the extent of the flood and the effect of the flood on terrain, vegetation, and city structures through sophisticated imaging systems and software.

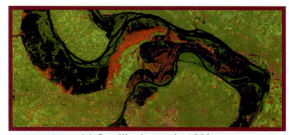

(a) Satellite image in 1993.

(b) Satellite image in 1994.

Figure 1.6. Satellite images of the St. Louis flood. (Image courtesy of NASA/Goddard Space Flight Center Scientific Visualization Studio.)

1.2.5 Robotics

The field of robotics has made astounding progress in recent years. Robots now appear on the shelves of commercial toy stores, in industrial manufacturing lines, and in search and rescue missions. At the heart of most intelligent robots is a set of image processing routines that is able to process images gathered by the robot's "eyes" and determine how the robots should respond to their visually perceived environment. A team of robotics experts from the University of Southern Florida was brought in to assist in the search and rescue mission during the days after the World Trade Center collapse. These robots were specifically designed to navigate through dangerous situations looking for signs of life.

1.2.6 Professional Sports

Most professional sports leagues are developing computer systems to improve either the sports telecast or to assist umpires and referees throughout the game. The US Tennis Association, for example, uses specialized image processing systems to assist in making line calls. Officials were having increased difficulty with making correct calls as skilled tennis players can now generate 150 mile-per-hour serves and 100 mile-per-hour backhands.

Major League Baseball has also installed complex image processing systems to record the trajectory of each pitch made during a baseball game. Two cameras track the motion of the ball and are able to triangulate the position to within 1/2 inch accuracy over the entire trajectory of the pitch. A third camera is used to monitor the batter and determine the strike zone by computing the batter's knee-to-chest position. While the system is not used during game play it is used to augment television broadcasts. High-performance image processing algorithms superimpose the pitch trajectory and strike zone on instant replays. This gives sports fans an objective way to decide if the pitch was a ball or a strike. Major League Baseball does use the system to rate the performance of plate umpires in calling balls and strikes. At the conclusion of each game, the plate umpire is given a CD-ROM containing the trajectories of every pitch along with a comparison between the computer and umpire calls made.

Other sports have successfully used image-processing techniques for both decision-making and aesthetic purposes. Most major networks airing National Football League games superimpose yellow "first down" markers onto the playing field. These yellow stripes are obviously not actually on the field, but are applied using real-time image processing techniques. With the decreasing cost of computational power, it is to be expected that image processing will become more prevalent in all areas of professional sports.

1.2.7   Astronomy

Astronomers have long used digital images to study deep space over much of
the electromagnetic spectrum: the Compton Gamma Ray Observatory captures
digital images primarily in the gamma ray spectrum; the Chandra X-Ray Observatory and the Space Infrared Telescope Facility (also known as the Spitzer
Space Telescope) provide coverage of the x-ray and infrared portions of the spectrum, respectively. The most well known telescope covering the visible portion
of the spectrum is the Hubble Space Telescope, which was launched in 1990.
The Hubble Telescope orbits the earth with a reflector-style optics system and a
mirror of 2.4 meters in diameter. The focal length is 57.6 meters and it is able
to take infrared images as well as images in the visible spectrum. Of course the
images are digital since they must be transmitted to ground stations for viewing
and analysis. The Hubble has produced some of the most remarkable images
ever taken of created order.

Figure 1.7 is an image of the Antennae galaxies. These two galaxies are
located in the constellation Corvus and are in the process of collapsing into a

Figure 1.7. Hubble Space Telescope image of the Antennae galaxies. (Image courtesy of
NASA, ESA, and the Hubble Heritage Team.)

single galaxy. These galaxies are approximately 45 million light years away, and scientists predict that within 400 million years the two galaxies will have merged to form a single elliptical galaxy.

1.2.8   Conclusion

Ours is an increasingly visual culture and digital imaging is pervasive across nearly all professions, disciplines, and academic fields of study. The study of digital image processing will provide a foundation for understanding how best to acquire digital images, the nature of information contained within a digital image, and how to best archive and display images for specific purposes or applications.

Artwork

Figure 1.3. *"Migrant Mother"* by Dorothea Lange (1895–1965). Dorothea Lange was born in Hoboken, New Jersey in 1895 and devoted herself to portrait photography at a young age. After apprenticing with a photographer in New York City, she moved to San Francisco and worked predominantly with the upper class. After about 13 years she developed the desire to see things from a different point of view and Lange began shooting among San Francisco's unemployed and documenting the increasing labor unrest. She was eventually hired by the Farm Security Administration (FSA) as a photographer and photojournalist. She is best known for her work with the FSA, which put a human face on the tragedy of the Great Depression and profoundly influenced the field of photojournalism in subsequent years. She died on October 11, 1965. Her most famous portrait is entitled "Migrant Mother," which is shown in Figure 1.3. The image is available from the United States Library of Congress's Prints and Photographs Division using the digital ID fsa.8b29516.

Optics and Human Vision

2

This chapter gives an overview of the physical properties of light, optics and the human visual system. The chapter provides important background for understanding how image processing mirrors the biological mechanisms of human perception and how the properties of human perception can be leveraged for computational advantage.

2.1 Light

In a physical sense light is composed of particles known as photons that act like waves. The two fundamental properties of light are the amplitude of the wave and its wavelength. Amplitude is a measure of the strength of the light wave where higher amplitude light waves are perceived as brighter. Wavelength measures, as the term itself indicates, the length of the wave and is typically given in meters. Light waves are also commonly characterized by their frequency, which is inversely proportional to the wavelength. The relationship between the wavelength λ and frequency f of a light wave traveling through a vacuum is

$$\lambda = \frac{c}{f},$$

where c is the speed of light, which is equal to 299,792,458 m/s. Since wavelength is given in meters, frequency is therefore given in units of seconds^{-1}, also known as hertz. Frequency measures the number of oscillations that a wave generates over a duration of one second of time. In terms of human perception, lower frequency (longer wavelength) light is perceived as the so-called *warmer* colors (red, yellow) while higher frequency (shorter wavelength) light is perceived as the *cooler* colors (violet, blue). While wavelength corresponds to color perception, amplitude corresponds to brightness, since brightness is proportional to the average energy over some time period. Figure 2.1 depicts a sinusoidal wave showing the relationship of amplitude and wavelength.

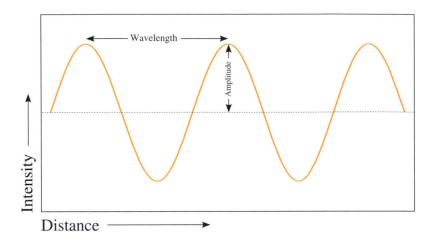

Figure 2.1. Diagram of a light wave.

The human eye can only perceive energy waves that have wavelengths within a very narrow band. Visible light is a relatively small portion of the larger electromagnetic spectrum which is composed of energy waves outside of our ability to perceive. The electromagnetic spectrum includes wavelengths as short as approximately 10^{-34} meters and as long as the size of the universe itself, although these are theoretical limits that have never been empirically verified. Electromagnetic radiation with wavelengths too small for the human eye to see are known, in decreasing order, as ultraviolet (UV), x-rays, and gamma rays. Electromagnetic radiation with wavelengths too large for the human eye to see are, in increasing order, infrared (IR), microwaves, and radio/TV.

Some living creatures are able to detect light outside of the range of wavelengths perceptible by humans. Bees, for example, are able to see UV radiation and pit viper snakes can see into the IR region using sensitive biological sensors in the pits of their heads. For human perception, however, the visible portion of the spectrum includes light with wavelengths ranging between about 400 and 700 nanometers. Light that has a wavelength of 650 nm appears red, while light that has a wavelength of 550 nm appears green; and light with a wavelength of about 475 nm appears blue. Figure 2.2 depicts the electromagnetic spectrum, highlighting the relatively narrow band of energy which is visible to the human eye. This figure depicts the portion of the spectrum spanning gamma rays, which have wavelengths on the order of 10^{-14} meters in length, to radio and television signals, which have wavelengths on the order of 10^4 meters.

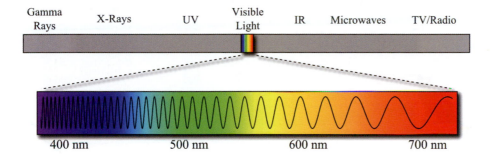

Figure 2.2. Electromagnetic spectrum.

2.2 Camera Optics

A lens is a transparent device, usually glass, that allows light to pass through while causing it to either converge or diverge. Assume that a camera is focused upon a target object. If the distance from the target to the lens is S_1 and the distance from the lens to the film is S_2 then these values are related by the thin lens equation shown in Equation (2.1), where f is the focal length:

$$\frac{1}{S_1} + \frac{1}{S_2} = \frac{1}{f}. \tag{2.1}$$

The focal length is a measure of how strongly a lens converges light and is defined as the distance from the lens to the point at which parallel rays passing through the lens converge. According to the thin lens equation, if we assume that an object is positioned a fixed distance S_1 from the lens, then a lens having a short focal length must produce correspondingly smaller images than a lens with a larger focal length. This follows since the focal length is directly proportional to the size of the image. Figure 2.3 illustrates light originating from a real-world target object, and passing through the lens and onto the imaging plane as a real image.

The magnification factor m of a lens is another measure of how strongly a lens converges light and is given by Equation (2.2). This formulation may initially seem inverted since digital cameras are often characterized by *optical zoom*, which is often misunderstood to be the magnification factor. The optical system of a camera is composed of multiple lenses where the positions of the lenses can vary and thereby adjust the focal length of a camera within some narrow range. The amount by which the focal length can be changed, and hence the resulting image size, is described as the *optical zoom*. The magnification factor

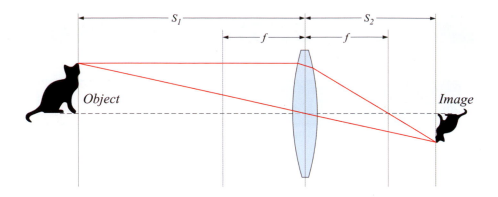

Figure 2.3. Optical properties of a convex lens.

of a single thin lens is typically less than 1 and should be understood to mean that an object of size S_1 appears to be of size S_2 when projected onto the focal plane:

$$m = \frac{\text{image size}}{\text{object size}} = \frac{S_2}{S_1}. \tag{2.2}$$

The aperture of a camera is the opening that determines the amount of light that is focused onto the imaging plane. The aperture of a camera is typically controlled by an adjustable diaphragm, which is allowed to expand or contract to either increase or decrease the amount of light entering the system. Of course the aperture can be no larger than the lens itself and is generally set smaller than the lens by the diaphragm. In photography the lens aperture is usually specified as an *f-number*, which is the ratio of focal length to the diameter of the aperture and is denoted as $f/\#$. A lens typically has preset aperture openings named *f-stops*, which allow a predetermined amount of light into the camera. These f-stops are conventionally set such that each successive f-stop either halves or doubles the total amount of light passing through the lens.

A circle of diameter d has an area of $\pi \cdot (\frac{d}{2})^2$, which implies that a halving of the area requires decreasing the diameter by a factor of $\sqrt{2}$. For this reason, the conventional f-stop scale is a geometric sequence involving powers of $\sqrt{2}$ as in the sequence $\{1, 1.4, 2, 2.8, 4, 5.6, \ldots\}$. Figure 2.4 illustrates how the diaphragm of the lens controls the amount of light passing through the optical system and shows the significance of the f-number. The open area of the lens labeled as $f/1.4$ is twice the area of the opening labeled as $f/2$, which is itself twice the area of the opening labeled as $f/2.8$.

Figure 2.4. F stops.

2.3 Human Visual System

2.3.1 The Human Eye

The human visual system is an exceptionally sophisticated biological imaging system composed of many interrelated biomechanical and biochemical sub-components. The eye is, of course, the primary component in this system and is itself composed of various parts. The eye is spherical in shape, having a diameter of approximately 20 mm. Figure 2.5 shows a cross section of the human eye and labels each of the primary structural components.

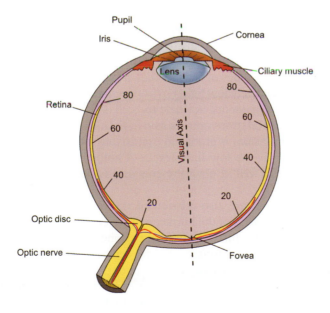

Figure 2.5. Primary structural elements of the human eye.

The cornea is a tough, transparent, dome-shaped membrane covering the front of the eye. The cornea is a protective covering over the eye and also functions to roughly focus incoming light. The iris, a diaphragm that expands and contracts to control the amount of light entering the eye, lies immediately behind the cornea. The central hole of the iris is the pupil, which varies in size from about 2 to 8 mm.

The lens is a convex and highly malleable disk positioned immediately behind the iris. While the cornea is responsible for most of the overall refraction of the eye, the curvature of the cornea is fixed, which requires the lens to fine-tune the focus of the eye. Ciliary muscles are used to adjust the thickness of the lens which adjusts both the curvature and the degree of refraction of the lens. When focusing at a distance, the lens flattens and therefore diminishes the degree of refraction while the lens thickens to focus on nearby objects.

The rear portion of the eye, known as the retina, is analogous to film in a conventional camera. The retina covers approximately 72% of the interior surface of the eye and is covered with two types of photoreceptor cells known as rods and cones. These biological sensors are able to convert light energy into electrical impulses, which are then transmitted to the brain through the optic nerve and interpreted as colors and images. There are approximately 6 million cones and 120 million rods in the human eye [Oyster 99]. The visual axis of the eye can be defined as a line extending between the center of the lens and the center of the retina, known as the fovea. The angle between any point on the retinal surface and the visual axis is given as the perimetric angle. Figure 2.5 shows the visual axis and the perimetric angles of various points on the retinal surface, spanning approximately $-80°$ to $+80°$.

Rods and cones are types of sensors that serve different purposes. Rods are able to function in relatively low light environments while cones function in relatively bright light environments. A single rod cell is able to detect a single photon of light and is about 100 times more sensitive than a cone cell. Scotopic vision, also known as night vision, is produced exclusively through rod cells since cone cells are not responsive to low light levels. Photopic vision is the normal vision of the eye under well-lit environments and is produced through the cone cells.

The rods and cones are also distributed throughout the retina in different ways. The 120 million rods of the human eye are concentrated around the outer edges of the retina and are therefore used in peripheral vision, while the 6 million cones in the retina are concentrated near the center of the retina, becoming less dense near the outer ring. This implies that in dark conditions the eye can more easily see objects by focusing slightly off-center of the object of interest, while in well-lit conditions the eye can more easily see by focusing directly on the target. Figure 2.6 plots the distribution of the rods and cones as a function of perimetric angle. Whereas the cones are clustered near the fovea (at a perimetric angle of $0°$), there are essentially no rods. As the perimetric angle moves away from

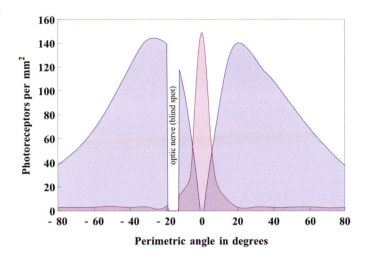

Figure 2.6. Photoreceptor density in the human eye as a function of perimetric angle.

the visual axis, however, the density of cones decreases sharply while the rod density increases sharply, peaking at approximately $\pm 35°$.

The optic nerve is responsible for carrying signals from the photoreceptors of the retina to the brain for analysis and interpretation. The optic nerve is composed of 1.2 million synaptic fibers and introduces a blind spot in the human eye since there are no photoreceptors in that area of the retina. Figure 2.6 indicates that the blind spot is at a perimetric angle of about 15° on the nasal periphery (the hemisphere of the eye oriented toward the nose).

Perhaps the most important distinction to make between the rods and cones is the role each has in discerning color. The rods, which are operative in scotopic vision, have no ability to discern color and hence see only in shades of gray. The primary reason that objects look washed out in dim light is that only the rods are active and can only detect differences in light intensity, not differences in color. Cones, by contrast, are able to discern differences in both intensity and color. The ability to see color derives from the fact that there are three separate types of cones. The three cone types respond differently to different light wavelengths such that one cone is sensitive to red, another to green, and another to blue wavelengths of light. These three types of cone are known as **L**, **M**, and **S** since they respond to **l**ong, **m**edium and **s**hort wavelengths of light, respectively. The **L** type cone has a peak sensitivity at wavelengths near 564–580 nm (red) while the **M** type peaks at 534–545 nm (green) and the **S** type peaks at 420–440 nm (blue). Color perception is thus gained by combining the sensory information of the three cone types. Roughly 65% of all cones are **L** type while 30% are **M** and 5% are **S** [Roorda et al. 01]. Table 2.1 is derived from [Kandel et al. 00] and summarizes the differences between the rods and cones.

Rods	Cones
Used for night vision	Used for day vision
Loss causes night blindness	Loss causes legal blindness
Low spatial resolution with higher noise	High spatial resolution with lower noise
Not present in the fovea	Concentrated in the fovea
Slower response to light	Quicker response to light
One type of photosensitive pigment	Three types of photosensitive pigment
Emphasis on detecting motion	Emphasis on detecting fine details

Table 2.1. Comparison between rods and cones.

2.3.2   Log Compression

Experimental results show that the relationship between the *perceived* amount of light and the *actual* amount of light in a scene are generally related logarithmically. The human visual system perceives brightness as the logarithm of the actual light intensity and interprets the image accordingly. Consider, for example, a bright light source that is approximately six times brighter than another. The eye will perceive the brighter light as being approximately twice the brightness of the darker. Log compression helps to explain how the human visual system is able to sense light across an exceptionally broad range of intensity levels.

2.3.3   Brightness Adaptation

The human visual system has the remarkable capacity of seeing light over a tremendously large range of intensities. The difference between the least amount of light needed to see (the scotopic threshold) and the maximum amount of light that we can see (the glare limit) is on the order of 10^{10}. The key to understanding how it is possible to perceive images across such a vast range of light levels is to understand that we cannot perceive light at both the low and upper limits *simultaneously*. At any particular instant we can discriminate among a very small portion of our full dynamic range. The sensitivity of the eye changes, or adapts, to the average brightness of the region on which it is focused. The eye calibrates itself to the average intensity of the region on which the eye is focused; it is then able to discriminate among a few dozen light levels which are centered about the calibration point. The range of intensities that the eye can see at one point in time is known as the instantaneous range of vision, while the calibration or adjustment to ambient lighting conditions is known as brightness adaptation.

Figure 2.7 shows that perceived brightness is a nonlinear function of the actual light intensity. The eye can perceive light across the full adaptive range but at any point in time can see only within the much narrower instantaneous range. The adaptive range is on the order of 10^{10} units and hence the graph vertical axis is shown using a log-compressed scale.

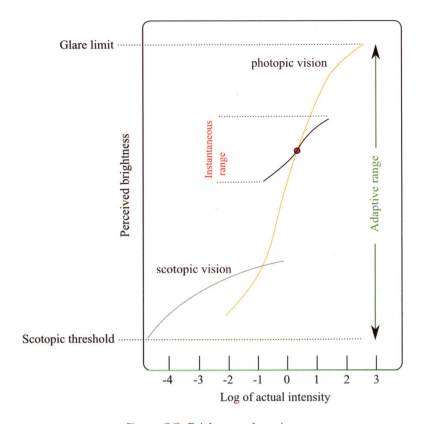

Figure 2.7. Brightness adaptation.

Consider taking a walk around noon on a cloudless and bright summer day. While the light is bright you are nonetheless able to discern with clarity your surroundings and make out details of the flowers and faces that you encounter. As you return home, however, and step through the doorway into a dark hallway entrance, the image is dull and dark. For a few moments the eye perceives almost pure darkness but as your visual system adapts to the overall brightness of the interior room, the sensitivity increases and more details become clear and visible. After some time it may even appear as if the interior room is as well lit as the sunny outdoors. While brightness adaptation allows us to see in extreme lighting conditions it has the disadvantage that we are unable to accurately estimate the actual intensity of light in our environment.

High dynamic range imaging (HDR) is an image processing technique that gives a displayed image the appearance of greater contrast than is actually displayed. HDR works by taking advantage of local brightness adaptation to fool the eye into thinking that it is seeing a much larger range of values than it really is.

2.3.4   Mach Banding

When viewing any scene the eye rapidly scans across the field of view while coming to momentary rest at each point of particular interest. At each of these points the eye adapts to the average brightness of the local region immediately surrounding the point of interest. This phenomena is known as *local* brightness adaptation. Mach banding is a visual effect that results, in part, from local brightness adaptation.

Figure 2.8 gives a popular illustration of Mach banding. While the figure of part (a) contains 10 bands of solid intensity the eye perceives each band as being either brighter or darker near the band edges. These changes in brightness are due to the eye scanning across the edge boundary, thus causing local adaptation to occur; this results in a changing perception of light intensity in that region when compared to the centers of the bands. The graph of part (b) is an approximation to perceived brightness as the visual system sharpens edges by increasing the contrast at object boundaries.

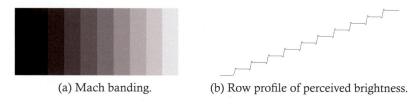

(a) Mach banding.   (b) Row profile of perceived brightness.

Figure 2.8. Mach banding effect.

2.3.5   Simultaneous Contrast

Simultaneous contrast refers to the way in which two adjacent intensities (or colors) affect each other. The image of Figure 2.9 is a common way of illustrating that the perceived intensity of a region is dependent upon the contrast of the region with its local background. In this figure, the four inner squares are of

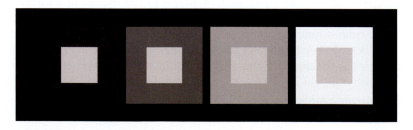

Figure 2.9. Simultaneous contrast.

identical intensity but are contextualized by the four surrounding squares, and thus the perceived intensity of the inner squares varies from bright on the left to dark on the right. This phenomena can be intuitively explained by noting that a blank sheet of paper may appear white when placed on a desktop but may appear black when used to shield the eyes against the sun [Gonzalez and Woods 92].

2.4   Exercises

1. A professional baseball league intends to save money and increase accuracy by replacing all umpires by automated computer vision systems. Miller Park will participate in a test run of the system and is purchasing a set of digital cameras that will serve as the front-end to the system. Each camera must be able to view a baseball at a resolution of at least 75 pixels per inch (75 digital pixels per inch of actual baseball) from a distance of at most 90 feet. Specify the minimum magnification and focal length required if the camera uses a CCD chip having 2048×2048 square pixels in an active sensor area of 3.55×3.55 mm. Clearly state any assumptions that you make concerning your design.

2. Section 2.3.2 states that perception of light intensity is logarithmically related to the actual light intensity. Empirical studies have shown that a logarithmic base of approximately 2.5 should be used to model this logarithmic compression when viewing a point light source. Using this model, what is the perceived intensity of a point light source emitting 120 units of light intensity?

3. A graphic artist wants to construct an image similar to that shown in Figure 2.8(a) but which contains no Mach banding. Discuss how such an image could be constructed and provide a convincing argument that your image minimizes the Mach banding effect.

Digital Images 3

3.1 Introduction

A digital image is visual information that is represented in digital form. This chapter describes common mathematical models of color and also gives various techniques for representing visual information in digital form.

3.2 Color

As described earlier, the human visual system perceives color through three types of biological photo-sensors known as cones. Each cone is attuned to one of three wavelengths that correspond roughly to red, green, or blue light. The L-cone, or long wavelength cone, responds to red light, the M-cone, or medium wavelength cone, responds to green light, and the S-cone, or short wavelength cone, responds to blue light. The individual responses of all cones within a small region of the retina are then combined to form the perception of a single color at a single point within the field of view. The design of this biological system suggests that color is a three-dimensional entity.

A color model is an abstract mathematical system for representing colors. Since color is a three dimensional entity, a color model defines three primary colors (corresponding to three dimensions or axes) from which all possible colors are derived by mixing together various amounts of these primaries. Color models are typically limited in the range of colors they can produce and hence represent only a portion of the visible spectrum. The range of colors covered by a color model is referred to as either the gamut or the color space.

Color models can be classified as either an additive or subtractive. Additive color models assume that light is used to generate colors for display. In an additive color model, the color black represents a complete lack of the primary colors while the color white corresponds to maximal and equal amounts of each of the primaries. Additive color models assume that the individual primaries sum together into a single color. Human perception is an additive system since the perception of black is caused by the complete absence of light while white is perceived when large and equal amounts of red, green, and blue are present. Human perception is often described as a *tristimulus color space*, since it is based on an additive color model composed of three primaries. Computer monitors and LCD projectors are light-emitting devices that are naturally modeled using an additive color model.

Subtractive color models assume that pigment will be used to create colors such that a complete absence of any pigment corresponds to the color white while combining the three primaries in maximal and equal amounts yields the color black. Subtractive color models tend to be the more intuitive of these two ways of defining color since people are generally accustomed to this mode; since they tend to use pencil, ink, or paint pigments to create images. Subtractive color models assume that when white light is projected onto pigment, the pigment will absorb power from certain wavelengths and reflect power at other wavelengths. This absorption is described as *subtraction* since certain wavelengths are subtracted from any light that is projected onto the pigment. In terms of electronic systems, images rendered with ink-based printers are most naturally described using a subtractive color model.

Numerous color models are in common use, each having qualities that are well suited for a certain class of applications. These include the RGB, CMY, CMYK, HSB, and NTSC (or YIQ) models. Since matching the characteristics of a particular color model to the requirements of a particular application is an important task in the design of an image processing system, it is important to understand how color models are designed along with their properties. The following subsections discuss the most commonly used color models for image processing applications and describe their basic properties in greater detail.

3.2.1   RGB Color Model

The RGB color model is the most common way of representing color in image processing systems. The RGB model is additive and uses red, green, and blue as the primary colors or primary axes such that any color can be obtained by combining different amounts of these three primaries. By way of example, consider a flashlight that has a slider allowing one to choose the strength of light emitted. In setting the slider to zero, the flashlight is turned completely off and generates no light; in setting the slider to one, the flashlight generates as much light as it is capable of generating. Now consider three such flashlights: the first emits purely red light, the second emits purely green light, and the third emits purely blue light. If all three flashlights are aimed at the same spot on a white wall any color can be projected onto the wall by adjusting the slider values on the three lights

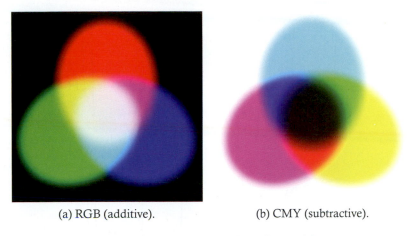

(a) RGB (additive).   (b) CMY (subtractive).

Figure 3.1. Additive and subtractive models.

in different ways. If all sliders are set to zero, black is projected onto the wall. If all sliders are set to one, white is projected onto the wall, and if all sliders are set to 0.5, then gray is projected.

Figure 3.1 illustrates the RGB color model with three lights projecting onto a white wall in an otherwise black room. The location at which they overlap indicates the sum of all three lights which is seen as white. The secondary colors are produced by the overlap of only two of the three primaries. Figure 3.1 also shows the CMY model which is described in the following subsection.

The system of three flashlights is a precise analog of the RGB color model. A color within the RGB color space is defined by three numeric values (a *tuple*) that specify the amount of red, green, and blue that comprise the specified color. A color specified using the RGB model is said to lie within the RGB color space. In this text we adopt the convention that an RGB color specification is specified in normalized coordinates such that the value 0 indicates that none of the specified primary is present and the value 1 indicates that the maximum amount of the specified primary is present. For example, the color red is specified in RGB space as the tuple $\langle 1, 0, 0 \rangle$, the color cyan is specified as $\langle 0, 1, 1 \rangle$, and middle gray is specified as $\langle .5, .5, .5 \rangle$.

The RGB color space can be visualized as the unit cube shown in Figure 3.2. All colors representable in RGB space lie within the volume defined by the three primary axes corresponding to pure red, green, and blue. The origin lies at $\langle 0, 0, 0 \rangle$ (black) while the opposite corner lies at $\langle 1, 1, 1 \rangle$ and corresponds to the color white. Each point within the RGB color space corresponds to a unique color as illustrated in Figure 3.2(b). Note that the line connecting pure black and pure white within the RGB color space is known as the gray scale, as shown in Figure 3.2(c) since any point lying on that line is a shade of gray.

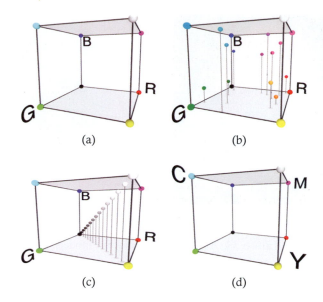

Figure 3.2. RGB and CMY color space.

Since RGB is an additive color model, the most common use of this model is for the display of images on computer monitors or projectors. A computer monitor is comprised of tightly packed red, green, or blue lights with varying degrees of intensity. Three such dots constitute a single point of color when viewed at a reasonable distance and hence the use of the additive RGB model is a natural choice for representing color in such a system. Also, since most digital images are meant to be displayed on a monitor, the RGB model is often used throughout the image processing pipeline as the data moves from acquisition to display.

3.2.2 CMY Color Model

Just as red, green, and blue are the primary colors of light, cyan, magenta and yellow are the primary colors of pigment. The CMY color space can, like RGB, be viewed as a cube where a CMY color is designated by a normalized tuple of values. Unlike RGB, however, the CMY model is subtractive. By way of example, consider an artist who paints landscapes using oil-based paint on a white canvas. Perhaps the artist is poor and can only afford to purchase three different pigments of oil paint. The artist wisely decides to purchase cyan, magenta, and yellow. The artist can still create any visible color by mixing together these three pigments in various proportions such that black can be created by mixing together equal amounts of all three colors while white can be created by mixing none of them together (i.e., leaving the white canvas empty). The CMY color model is illustrated in Figure 3.2(d) and is inversely related to the RGB color model. Given a color in RGB space as the tuple $\langle R, G, B \rangle$, the same color would be given in CMY space as the tuple $\langle C, M, Y \rangle$, where each of the three values are computed as shown in Equation (3.1).

$$
\begin{aligned}
C &= 1 - R, \\
M &= 1 - G, \\
Y &= 1 - B.
\end{aligned}
\tag{3.1}
$$

The three formulae of Equation (3.1) are typically written using matrix notation as

$$
\begin{bmatrix} C \\ M \\ Y \end{bmatrix} = \begin{bmatrix} 1 - R \\ 1 - G \\ 1 - B \end{bmatrix}.
$$

The CMY color model is most commonly used in the print industry since it directly corresponds to the use of pigment (ink) as a medium. Printers can generate color images from only three inks by mixing together various amounts of cyan, magenta, and yellow primaries at each point on the paper. The process of converting a color image into CMY layers is known as color separation.

The CMYK color model is an extension of the CMY. The CMYK model augments the CMY by including a fourth primary component that corresponds to the amount of black in a color. Even though the CMYK color space has an extra "axis," it does not contain more colors than CMY. In technical parlance we say that CMY and CMYK are *equally expressive* and hence mathematically equivalent. While CMYK is not more expressive than CMY, the additional parameter is nonetheless beneficial since it allows for the

♦ *Conservation of ink.* Many images have large black areas and it is wasteful to mix three pigments (CMY) together to form one black pixel.

♦ *Creation of black pixels.* According to the CMY model, mixing maximal amounts of cyan, magenta, and yellow will produce pure black. In actual practice, however, the result is a color that is muddy and dark but not truly black.

♦ *Creation of crisp black edges.* Mixing three colors to create black will not only yield a color not truly black, but will also result in dots and lines having blurred edges since the individual CMY color samples will not be precisely aligned with each other.

Conversion from CMY to CMYK is a nonlinear process. Assume that $\langle C', M', Y' \rangle$ is the CMY color that we wish to convert to CMYK. The CMYK values are then defined, as shown in Equation (3.2).

$$\langle C, M, Y, K \rangle = \begin{cases} \langle 0,0,0,1 \rangle & \text{if } \min(C', M', Y') = 1, \\ \langle \frac{C'-K}{1-K}, \frac{M'-K}{1-K}, \frac{Y'-K}{1-K}, K \rangle & \text{otherwise where } K = \min(C', M', Y'). \end{cases}$$

$$(3.2)$$

In order to convert from RGB to CMYK, first convert the RGB coordinates into CMY space using Equation (3.2.2) and then apply Equation (3.2).

3.2.3   HSB Color Model

The HSB color model decomposes color according to how it is *perceived* rather than, as is the case with RGB, with how it is *physically sensed*. Since HSB seeks to mirror the perception of color, the three dimensions of the HSB color space are aligned with a more intuitive understanding of color than either RGB or CMY. A point within the HSB gamut is defined by hue (the chromaticity or pure color), saturation (the vividness or dullness of the color) and brightness (the intensity of the color). The HSB color space can be visualized as a cylinder, as shown in Figure 3.3. The central vertical axis ranges from black at the bottom to white at the top. The most vivid colors lie on the external surface of the cones while duller colors occupy the interior portion where the central vertical axis of the cylinder is equivalent to grayscale. HSB is also known as HSI or HSL, using the terms *intensity* and *luminance* as synonyms for *brightness*. The HSV color model is also closely related to HSB but differs somewhat in the brightness band.

Hue is specified as an angular measure where red corresponds to $0°$, green corresponds to 120 degrees, and blue corresponds to $240°$. Hue is therefore cyclical and is well defined for any real value. If hue is normalized, as is most often the case, the value zero corresponds to 0 degrees (red) while the value 1 corresponds to $360°$ (and hence is equivalent to 0 since $0°$ and $360°$ are equivalent). Brightness is a normalized amount where 0 indicates no intensity (black) and 1 represents full intensity (white). The brightness setting selects a color wheel slice

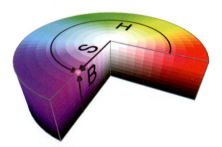

Figure 3.3. HSB color space.

perpendicular to the central vertical axis. Since saturation is an indication of vividness or colorfulness, saturation is defined as a percentage measure from the central vertical axis to the exterior shell of the cylinder. From this it should be clear that a saturation of 0 indicates a completely de-saturated or non-colorful grayscale value at the center of the cylinder while a saturation of 1 (or 100%) indicates a completely saturated and hence completely vivid color.

The color red is represented in HSB space as $\langle 0, 1, 1 \rangle$ and middle gray is represented as $\langle 0, 0, .5 \rangle$. The HSB model is degenerate at $B = 0$ since where there is no brightness the color must be black regardless of the hue and saturation. In other words, when the brightness value is set to 0 the values of H and S are irrelevant. The HSB model is also degenerate at $S = 0$ since the Hue value is then irrelevant.

Manipulating colors in HSB space tends to be more intuitive to artists than working with either the RGB or CMY models. In order to darken a color in HSB only the brightness value is reduced while darkening a color in RGB involves scaling each of the three primary components proportionally. The complement of a color in HSB is easily found by crossing to the opposite side of the HSB cylinder, computationally achieved by shifting the hue band by $180°$.

The HSB color model is often used in image processing since it provides a way to separate color information (HS) from intensity (B). This is useful since image processing applications often operate only on the intensity of an image while leaving the color unaltered. Many of the image processing operations in this text that are defined on grayscale images can be performed on color images by first converting the image to HSB, applying the grayscale operation to the brightness band only, and then converting back to the original color space.

HSB is a nonlinear transformation of the RGB color space. Let $\langle R, G, B \rangle$ be a normalized RGB color that is to be converted into HSB space. Let $Max = \max(R, G, B)$ and $Min = \min(R, G, B)$. Equation (3.3) then shows how to convert from normalized RGB space into normalized HSB space:

$$H = \begin{cases} \text{undefined} & \text{if max} = \text{min}, \\ 60 \times \frac{G-B}{\text{max} - \text{min}} & \text{if max} = R \text{ and } G \geq B, \\ 60 \times \frac{G-B}{\text{max} - \text{min}} + 360 & \text{if max} = R \text{ and } G < B, \\ 60 \times \frac{B-R}{\text{max} - \text{min}} + 120 & \text{if max} = G, \\ 60 \times \frac{R-G}{\text{max} - \text{min}} + 240 & \text{if max} = B; \end{cases} \tag{3.3}$$

$$S = \begin{cases} 0 & \text{if max} = 0, \\ 1 - \frac{\text{min}}{\text{max}} & \text{otherwise}; \end{cases}$$

$$B = \text{max}.$$

3.2.4   YIQ Color Model

Much like the HSB color model, the YIQ color model possesses a single grayscale axis and two axes that relate to color. The Y axis contains intensity information (grayscale values) while chromatic information is packed into the I and Q channels. The YIQ model can be thought of as a rotation and distortion of the RGB space such that the Y axis lies along the RGB grayscale diagonal. The two color axes then correspond roughly to red-green (I) and blue-yellow (Q).

The YIQ color model is of interest due largely to the way it has been adopted by broadcasters and audio/video electronics manufacturers. In the United States, for example, black and white television broadcasts during the mid 1900s used a single grayscale channel for image transmission. When the National Television Standards Committee (NTSC) developed the standard for *color* television broadcasts, they chose YIQ as the color model to allow backward compatibility with black and white television sets. Since the Y channel of the color standard is equivalent to the grayscale of the black and white standard, black and white television sets were able to tune into the Y channel of a color broadcast while discarding the I and Q channels. In addition, since human perception is far more sensitive to brightness than color, separation of color and brightness bands allows more information to be packed into the Y channel than the I and Q channels. The NTSC standard allocates 4 megahertz (MHz) to the Y channel, 1.5 MHz to the I channel, and 0.6 MHz to the Q channel where a MHz is a measure of the amount of information that can be carried on a channel. The standard essentially allocates 66% of a television signal for intensity, 25% for red/green, and the remaining 9% for blue/yellow.

An RGB color can be converted to YIQ space using the transformation of Equation (3.4). Note that the elements of the first row in the conversion matrix sum to 1 while the elements of the other rows sum to zero. Since all chromatic information is packed into the I and Q channels, we would expect that a grayscale RGB value would reduce to zero in both the I and Q dimensions. Since each of the R, G, and B samples of an RGB grayscale color are identical, it follows that the coefficients must themselves sum to zero. Also note, that while normalized RGB coordinates are in the range $[0, 1]$, the range of values for the I and Q bands includes negative values:

$$\begin{bmatrix} Y \\ I \\ Q \end{bmatrix} = \begin{bmatrix} 0.299 & 0.587 & 0.114 \\ 0.596 & -0.274 & -0.321 \\ 0.211 & -0.523 & 0.312 \end{bmatrix} \begin{bmatrix} R \\ G \\ B \end{bmatrix}. \tag{3.4}$$

3.2.5   YCbCr Color Model

The YCbCr color model possesses many of the same qualities as YIQ and is the color model used in the JPEG image file format as well as in component video systems. YCbCr is often confused with the YUV color model and it is most often the case that YUV should be understood as YCbCr. As in the YIQ space, luminance is isolated in the Y band while the Cb and Cr axes represent chromaticity. The Cb band is the difference between the blue component and a reference point while the Cr band is the difference between the red component and a reference point. The YCbCr coordinates can be computed from normalized RGB values, as shown in Equation (3.5).

$$
\begin{bmatrix} Y \\ Cb \\ Cr \end{bmatrix} = \begin{bmatrix} 16 \\ 128 \\ 128 \end{bmatrix} + \begin{bmatrix} 65.481 & 128.553 & 24.966 \\ -37.797 & -74.203 & 112.000 \\ 112.000 & -93.786 & -18.214 \end{bmatrix} \begin{bmatrix} R \\ G \\ B \end{bmatrix}. \quad (3.5)
$$

Note that the YCbCr coordinates are in 8-bit precision according to the transformation of Equation (3.5) and that the range of the various components does not fully cover 8 bits. The Y band is in the range $[16, 235]$ while the Cb and Cr bands are in the range $[16, 240]$. In image processing applications these bands may then need to be stretched (scaled) into the full 8-bit dynamic range.

3.2.6 Color Similarity

In many image processing applications it is necessary to determine how different one color is from another. Since colors are given as points in some three-dimensional color space, the difference between two colors can be defined as the distance between the two points. As the distance between two colors grows smaller, the two colors are more similar and of course as the distance increases the two colors are more dissimilar. Consider as an example the two RGB colors $C_1 = (r_1, g_1, b_1)$ and $C_2 = (r_2, g_2, b_2)$. We wish to compute the distance between C_1 and C_2 as a way of measuring their similarity. The distance between two three-dimensional points can be computed in different ways; the two most common metrics are known as the L_1 distance (also known as the Manhattan or taxi cab distance) and the L_2 distance (also known as the Euclidean distance). These two ways of measuring distance are defined is Equation (3.6).

$$
\begin{aligned}
L_1 &= |r_2 - r_1| + |g_2 - g_1| + |b_2 - b_1|, \\
L_2 &= \sqrt{(r_2 - r_1)^2 + (g_2 - g_1)^2 + (b_2 - b_1)^2}.
\end{aligned} \quad (3.6)
$$

The taxi cab metric derives its name from the notion that a taxi cab in any large metropolitan city must travel from point to point by navigating streets that are aligned on some grid. The Euclidean distance measures the straight-line distance between two points. Figure 3.4 gives a visual illustration of the taxi cab metric as applied to the RGB color space.

Distances between colors must often be normalized for computational purposes. The normalized distance between two colors C_1 and C_2 is determined by

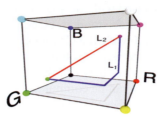

Figure 3.4. Taxi cab and Euclidean distance metrics.

first computing the distance between C_1 and C_2 and then dividing that distance by the maximum possible distance between any two colors in the given color space. The maximum distance between any two colors in RGB space, for example, occurs when the two colors are at diagonally opposite corners of the color cube. The color pairs black/white, red/cyan, green/magenta, and blue/yellow have maximal distances which are given as 3 for the L_1 metric and $\sqrt{3}$ for L_2.

As an example of color similarity, consider the normalized RGB colors $(.5, .4, .2)$ and $(.3, .7, .1)$. The taxicab distance between these colors is .6 and the Euclidean distance is .374, as shown in Equation (3.6). The normalized taxicab distance between these two colors is $.6/3 = .2$ and the normalized Euclidean distance is $.374/\sqrt{3} = .216$.

$$
\begin{aligned}
L_1 &= |.5 - .3| + |.4 - .7| + |.2 - .1| \\
&= .2 + .3 + .1 \\
&= .6, \\
L_2 &= \sqrt{(.5 - .3)^2 + (.4 - .7)^2 + (.2 - .1)^2} \\
&= \sqrt{(.04 + .09 + .01)} \\
&= \sqrt{.14} \\
&= .374.
\end{aligned}
$$

Note that the L_1 and L_2 distance can only be directly applied to those color models that are represented in some Cartesian coordinate space. RGB, CMY, YIQ, and YUV color models fall into this category while the HSB model does not. The HSB color model uses a cylindrical model space rather than Cartesian since hue is given as an angular measure and saturation as a radial distance from the center. In order to determine the distance between two HSB colors we must first convert them into some Cartesian space after which one of the metrics given in Equation (3.6) can be applied.

Given a normalized HSB color $C = (h, s, b)$ we can convert into a Cartesian coordinate space where the x-axis is aligned along the red direction of the HSB cylinder, the z-axis is aligned with brightness and the y-axis is perpendicular to both x and z. Projection of the HSB values onto these axes is given by Equation (3.7):

$$
\begin{aligned}
x &= s \cos(2\pi h), \\
y &= s \sin(2\pi h), \\
z &= b.
\end{aligned}
$$

The distances between two HSB colors (h_1, s_1, b_1) and (h_2, s_2, b_2) can then be computed by applying Equations (3.7) to obtain (x_1, y_1, z_1) and (x_2, y_2, z_2) after which Equation (3.6) can be directly applied, as shown in Equation (3.7). Note that the maximum distance between two colors in normalized HSB space is given as $1 + 2\sqrt{2}$ for the L_1 metric and $\sqrt{5}$ for L_2. Also note that the L_1 metric in HSB space is not consistent since the L_1 distance between two fully saturated complementary colors having the same brightness will range between 2 and $2 * \sqrt{2}$:

$$
L_1 = |s_1 \cos(2\pi h_1) - s_2 \cos(2\pi h_2)| + |s_1 \sin(2\pi h_1) - s_2 \sin(2\pi h_2)| + |b_1 - b_2|,
$$

$$
L_2 = \sqrt{(s_1 \cos(2\pi h_1) - s_2 \cos(2\pi h_2))^2 + (s_1 \sin(2\pi h_1) - s_2 \sin(2\pi h_2))^2 + (b_1 - b_2)^2}.
$$

3.2.7   Representing Color in Java

As the previous sections have indicated, color is a three dimensional entity. The three dimensions of color form a volumetric space wherein each point represents a color. Any representation of color must then include the storage of three values along with some indication of the color space used. The Java graphics library, `java.awt`, contains a class that does precisely this. The `Color` class represents a color by storing three numeric values along with an indication of the color space used. While the `Color` class is flexible enough to represent colors in any color space, Java is primarily used to create images for display on computer monitors and hence RGB is the default color space. A partial listing of methods in the `Color` class is given in Figure 3.5.

The `Color` class supports two constructors. A `Color` object can be constructed by providing a normalized RGB tuple (each floating point value is in the range $[0, 1]$) or an 8-bit tuple (each integer is in the range $[0, 255]$). While the two constructors listed above use the default RGB color space, there are additional methods not listed here that allow clients to specify and manipulate colors in alternative color spaces. Care should be taken, however, when using alternative color spaces since image display is optimized for images in RGB space and using other spaces may significantly degrade system performance. Regardless

Color
Color(float r, float g, float b)
Color(int r, int g, int b)
int getRed()
int getGreen()
int getBlue()

Figure 3.5. Color class methods (partial listing).

of the color space used, all Color objects are internally able to convert themselves into 8-bit RGB space and provide the corresponding values through the getRed(), getGreen(), and getBlue() methods.

Java does not provide any support for computing the similarity between two Color objects. As a result, we must author customized methods to compute the distance between any two colors. Consider the development of a class named ColorUtilities that contains a collection of static methods for performing basic color operations and transformations. Listing 3.1 gives a partial implementation of this class, including pairs of overloaded Java functions to compute the

```java
public class ColorUtilities {
  public static float euclideanDistance(float s01, float s11, float s12,
                                        float s02, float s12, float s22) {
    float db0 = (s01-s02);
    float db1 = (s11-s12);
    float db2 = (s21-s22);
    return Math.sqrt(db0*db0 + db1*db1 + db2*db2);
  }

  public static float euclideanDistance(Color c1, Color c2) {
    return euclideanDistance(c1.getRed(), c1.getGreen(), c1.getBlue(),
                  c2.getRed(), c2.getGreen(), c2.getBlue());
  }

  public static float taxiCabDistance(float s01, float s11, float s12,
                                      float s02, float s12, float s22) {
    return Math.abs(s01-s02) + Math.abs(s11-s12) + Math.abs(s21-s22);
  }

  public static float taxiCabDistance(Color c1, Color c2) {
    return taxiCabDistance(c1.getRed(), c1.getGreen(), c1.getBlue(),
                  c2.getRed(), c2.getGreen(), c2.getBlue());
  }
}
```

Listing 3.1. The ColorUtilities class.

L_1 and L_2 distance between any two colors. In each pair of methods, a color can be given as either a `Color` object or by three separate floats that represent a normalized color in some Cartesian color space.

Having described the fundamental properties of color along with Java's built-in manner of representing color through the `Color` class, we now seek to represent a digital image. Given our understanding of color it is tempting to conclude that an image should be represented as a two-dimensional array of `Color` objects such as given in Listing 3.2.

```
1    Color[][]  image = new  Color[600][400];
```

Listing 3.2. Naive image representation.

As we will see in the following section, however, there are a number of important reasons for not using this approach. In fact, when writing image processing applications the `Color` class should be used sparingly. Although it may at times present a convenient representation for color there are faster and more memory efficient techniques that are elaborated in the following section.

3.3   Digital Images

3.3.1   Image Types

Although all images can be thought of as full color images (even black and white are colors), it is useful to have a clear definition of terms that are commonly used to describe image types. The pixels of a color image are allowed to range over the full color space of some color model. In RGB, for example, a color image may contain pixels that fall anywhere within the color cube. The colors present in a grayscale image are restricted, however, to only those colors that lie on the black-white diagonal of the RGB color cube or, equivalently, on the central vertical axis of the YIQ or HSB color spaces. A grayscale image therefore has only intensity information and no chromaticity. Binary images, as the name implies, contain exactly two colors that are typically, but not necessarily, black and white. In this text, we recognize only three types or classes of image: color, grayscale, and binary. Figure 3.6 gives a visual example of these three image types.

Two other terms are commonly used to describe image types but they are often misused. A black and white image, for example, must contain exactly two colors (black and white) but is often incorrectly used to refer to a grayscale image. In addition, a monochromatic image must, by implication, be composed of a single color but is commonly used to refer to a binary image containing black pixels (the monochrome) on top of a white background.

(a) Color image.   (b) Grayscale image.   (c) Binary image.

Figure 3.6. Image types.

Since a digital image attempts to accurately represent an image in digital form, a conceptual model of the structure of an image is first required. The most common conceptual model of a digital image is that of a two-dimensional grid of colored picture elements or pixels. Each pixel corresponds to a small rectangular region of an image that has been filled with a single color. Each pixel therefore has an implicit width, height, and location within the digital image in addition to the color. When the grid of pixels is viewed at an appropriate distance, the individual pixels recede from view while collectively appearing as a continuously colored whole. Each pixel, since it is a color, is itself composed of one or more *samples*. A single pixel in an RGB image, for example, contains a red sample, a green sample and a blue sample. A *band* is a grid of samples where the number of bands in the image is equivalent to the number of samples in a single pixel. In a grayscale image, for example, there is only one band while in an RGB image there are three bands (the grid of red samples, green samples, and blue samples) and in a CMYK image there are four bands. The term band is synonymous with the term *channel* used in many textbooks. Since, however, Java uses the term band this text also adopts that usage in order to avoid conflict or confusion.

In order to adequately present these concepts throughout the text we will now define a concise notation for describing images, bands, pixels and samples. Given an image I, we denote an individual pixel of this image by $I(x, y)$ where x corresponds to the *column* and y corresponds to the *row* of the pixel in the image. In image processing it is very important to note that the *top* row of an image is row *zero* and row numbers increase from top to bottom. Since a pixel may be

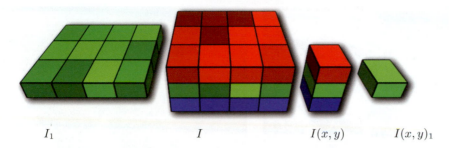

$$I_1 \qquad\qquad I \qquad\qquad I(x,y) \qquad I(x,y)_1$$

Figure 3.7. Structure of an RGB image.

composed of multiple samples, we denote the sample on band b of pixel $I(x,y)$ as $I(x,y)_b$. To access a single sample of an image then requires the use of three indices: two of which specify the pixel by column and row and one index which specifies the band. Figure 3.7 gives an example of the structure of an image in the RGB color space. The image I is composed of three bands: the red (top), green (middle), and blue (bottom). The green band is denoted as I_1 and is composed of all green samples of the the image. A single pixel is given by $I(x,y)$ while the green sample of that pixel is given by $I(x,y)_1$.

3.3.2 Color Depth

The *color depth* of an image describes the ability of the digital image to accurately reproduce colors. Color depth is given as the number of *bits* consumed by a single pixel of an image. A binary image for example, will have a color depth of 1 bit per pixel, or 1 bpp, since there are only two colors in the entire image and a single bit is sufficient to distinguish between them. A grayscale image will usually possess a color depth of 8 bpp and a color image 24 bpp. In a 24 bpp RGB color image, 8 bits of each pixel are allocated to each of the red, green, and blue bands. Each pixel is therefore able to represent one of 2^{24} colors (or $2^8 \cdot 2^8 \cdot 2^8$) or $16,777,216$ separate colors.

In terms of computation there is always a tradeoff between the color fidelity of an image and the resources required. While images with greater color depth are able to more accurately reproduce colors across the full spectrum they consume more bits (memory) than images of lesser color depth. A color depth of 24 bpp suffices for the vast majority of image processing applications since human visual perception is unable to distinguish between color changes that are smaller than what the 24 bpp depth supports. While images having a color depth exceeding 24 bpp are rare, certain specialized applications require higher color resolution. High quality scanners, medical imaging systems, and specialized scientific instruments often store images using color depths of 30 bpp. Figure 3.8 illustrates the effect of color depth on the quality of a digital image.

(a) 1 bpp color depth.   (b) 2 bpp color depth.

(c) 5 bpp color depth.   (d) 24 bpp color depth.

Figure 3.8. The effect of color depth on image quality.

(a) 4×5 resolution.

(b) 20×25 resolution.

(c) 100×125 resolution.

(d) 360×450 resolution.

Figure 3.9. The effect of resolution on image quality.

3.3.3   Resolution and Aspect Ratio

The *pixel resolution* of an image measures the amount of visual detail that an image holds where higher resolution means a greater capacity for visual detail. Image resolution in digital imaging is conventionally given by the number of columns and rows in the image. If, for example, an image has a resolution of 640×480 (read as *640 by 480*) the image has 640 columns and 480 rows for a total of 307,200 pixels. It is also popular to cite resolution by giving only the total pixel count, usually specified in terms of megapixels.

Pixel resolution often includes some notion of the physical size of the image when either acquired or displayed. When physical size is considered, the pixel resolution is given in terms of pixels per unit of length or pixels per unit of area. Printers, for example, may be characterized as 600 dots (pixels) per inch and geographic information systems often give pixel resolution as the ground sample distance (GSD), which is a measure of how much of the earth's surface area is covered by a single pixel of a satellite image.

The *aspect ratio* is a measure of the relative height and width of a digital image and is defined as the width of the image divided by the height of the image. The aspect ratio is often given in simplified terms so that for an image having a resolution of 640×480 the aspect ratio can be given as either 4:3 (read "four to three") or as 1.33:1 or as merely 1.33. Also, when a digital image is sampled the individual photosites have a width and a height which implies that individual pixels may have an aspect ratio. The pixel-level aspect ratio is usually assumed to be 1, but if the pixels of an image are not square then the actual aspect ratio of the image must be adjusted by multiplication with the pixel aspect. Figure 3.9 illustrates the effect of image resolution on image quality by allowing the pixel resolution to vary while holding other image attributes constant.

3.3.4   Frequency

The pixels within an image form patterns of varying intensities that are often described in terms of *frequency*. The frequency of an image is a measure of the amount of change in color that occurs over spatial distance, which is usually measured in terms of pixels. Regions of an image that contain large color change over a small distance are characterized as high frequency regions. Regions that have low color variance over large spatial distances are characterized as low frequency regions. The concepts of image frequency and resolution are directly related. High resolution images are able to capture high frequency details while low resolution images cannot adequately capture fast changes in color over short spatial distances.

Most images contain both high and low frequency regions as illustrated in Figure 3.10. In this figure, the region labeled **A** is a very low frequency region since the color varies very little across the pixels within the region. Region **B** is

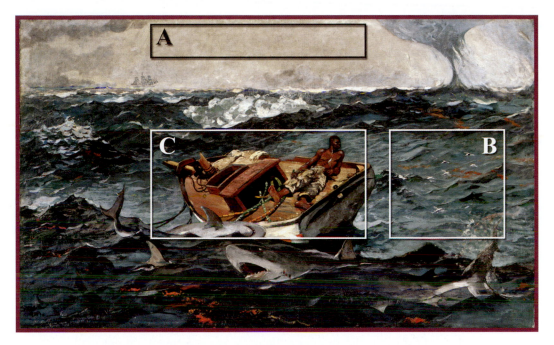

Figure 3.10. Image frequency.

characterized by relatively moderate frequency since the color variance is moderate over a smaller number of pixels, while region **C** can be characterized as a relatively high frequency region of the image. Analyzing image content in terms of frequencies is fundamental to much of image processing and is further detailed in Chapter 9.

In summary then, a digital image can be characterized by its resolution, color depth, and color model in addition to frequency content. Since each of these attributes affects image quality, careful selection of resolution, color depth, and color model are crucial when developing an image processing system.

3.4   Acquisition

3.4.1   Grayscale Images

Digital images can be acquired directly through the use of imaging hardware or they may be synthesized by software applications that allow a user to "draw" a digital image directly on a computer monitor. In the case of hardware-based acquisition systems, a real-world object or environment, referred to here as a continuous tone natural scene, is converted into a digital image through *sampling* and *quantization*. Sampling a continuous tone scene partitions the scene spatially into a grid of rectangular regions (the pixels) while quantizing the scene partitions the continuous range of light intensity levels into a discrete set of values.

Hardware-based acquisition systems such as digital cameras, digital scanners, and digital copy machines, rely on an electronic device known as an image sensor to perform both sampling and quantization. The two most common types of image sensors are based on two different technologies. While the charge coupled device (CCD) and the complementary metal oxide semiconductor (CMOS) sensors have slightly different approaches to image sensing, they are roughly equivalent in terms of function and produce images of comparable quality.

In either case, an image sensor is an electronic circuit with a grid of small, usually rectangular, photocells mounted on an electronic chip. The optical lenses of the system focus a scene onto the surface of the photo-sensors such that when an image is acquired, each photocell measures the total amount of light that falls within its boundaries. The light intensity of each sensor is recorded and, when viewed as a collective whole, the records represent a digital image.

When acquiring an image via an image sensor, the resulting digital image is of course an approximation to the continuous tone natural scene which has been sampled and quantized by the sensor. Sampling refers to the notion that a single pixel reduces an *area of light* into a *point* of light and hence a pixel should be understood as the amount of light present at a singular point within a scene. Quantization refers to how finely a sensor is able to measure the intensity of light. Quantization measures a continuous value, light intensity, and maps it into one of several possible digital values through a process known as analog-to-digital (A/D) conversion. Image sampling is directly related to image resolution where a higher sampling rate implies a greater image resolution. Image quantization is directly related to color depth such that better quantization implies greater color depth.

It is important to recognize that a sample represents a geometric point in the image plane and hence has no shape or size. Although the sample was likely acquired by a rectangular photo-sensor with nonzero width and height, the sample value itself represents the amount of light striking at the center of the photosite.

Image sampling and quantization are illustrated in Figure 3.11. In this example the continuous tone scene of part (a) is projected onto a CCD containing a 15 column by 17 row grid of photo-sensors. The physical size and shape of the photo-sensors serve to sample the source image and produce the digital image of part (b) which has a resolution of 15×17. The sampled image is then quantized by the CCD photo-sensors using an 8 bit color depth to produce the digital image of part (c). Since each pixel outputs a single numeric value, the output of the CCD is best understood as a grayscale image, even if the source image is color.

In this text, a real-world scene will be denoted as a continuous tone image. A continuous tone image can be modeled as a continuous function $f(x, y)$ over a two-dimensional space known as the image plane. The parameters x and y

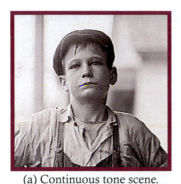 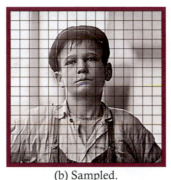 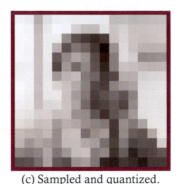

| (a) Continuous tone scene. | (b) Sampled. | (c) Sampled and quantized. |

Figure 3.11. Image sampling and quantization.

are Cartesian coordinates that specify a point in the image plane and the value of the function represents the amount of light present at the specified location. In human vision, the image plane corresponds roughly to the retina, the x-y coordinates locate points on the retina and the function f specifies the color and intensity of the light at the specified location. Modeling images as two dimensional functions of light is of interest to theoreticians since such a model is precise and subject to functional and symbolic analysis. Continuous models are rarely used in practice, however, due to the inherent difficulty of specifying and manipulating equations that correspond to visual data. As a result, digital images are almost always modeled as discrete approximations to some continuous analog image.

An individual pixel value corresponds to the total amount of light that strikes the rectangular area represented by the pixel itself. Consider, for example, a scenario where a digital camera in a dark cavern is aimed at a weak flashlight. Figure 3.12 gives a mathematical illustration of this scenario and highlights in more mathematical terms the process of sampling and quantization. In this figure, part (a) shows a plot of the continuous Gaussian distribution function $f(x, y) = e(x * x + y * y)$ over the interval $[-5, 5]$ in the x-y image plane. The image sensor lies in the x-y plane and the height of the function corresponds to the amount of light falling across the image sensor. The image of part (b) shows a projection of the individual photo sensors onto the plot and indicates how the function will be sampled. For this example, the pixel elements are $\frac{1}{2} \times \frac{1}{2}$ unit squares in size. The image of part (c) shows the effects of quantization where the light intensity values have been partitioned into eight possible values corresponding to a 3-bit color depth. The resulting digital image is shown in part (d), where white pixels correspond to regions of high light level while darker pixels correspond to regions of low light levels.

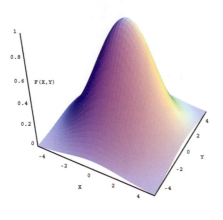

(a) Plot of a continuous Gaussian distribution function.

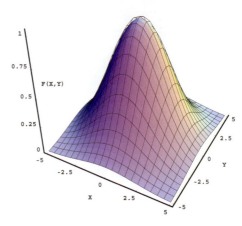

(b) Projection of individual photo-sensors.

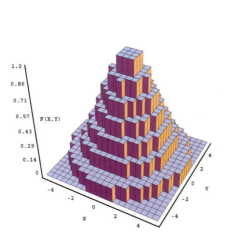

(c) Effects of quantization.

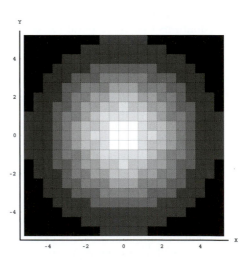

(d) Digital image.

Figure 3.12. Sampling and quantization.

3.4.2   Acquisition of Color Images

The most common technique for acquiring color images uses a single image sensor where each pixel site is filtered to sense either red, green, or blue light. The specific arrangement of red, blue and green elements is determined by a color filtering scheme. Since human vision is more sensitive to green wavelengths than red or blue, the most popular color filtering scheme is the Bayer filter, a patented technique named after its discoverer, Dr. Bryce E. Bayer. Every 2×2 sub-region of a Bayer filtered CCD contains one red, one blue, and two green pixels that roughly mimics the relative sensitivity of cones in human perception.

When an image is acquired using a Bayer-filtered sensor, the unprocessed result is a set of three images corresponding to the red, green, and blue bands such that each of the bands contains gaps. This data is commonly known as the *RAW* output. The RAW data must be processed by combining the three partially complete bands in such a way as to produce a seamless true-color image. The process of combining the three bands of RAW data is known as *demosaicing*. This process is illustrated in Figure 3.13, where the Bayer filtered CCD of part (a) is used to acquire the three bands shown in part (b). The RAW data of part (b) must then be demosaiced to form a color image that typically has a lower resolution than that of the image sensor itself. Bayer filtering generally reduces the resolution of the sensor in order to gain color depth.

An alternative to color imaging with Bayer filters is known as *3CCD*, a technique that, as the name implies, uses three separate image sensors. The optics in a 3CCD system split the incoming scene into three separate filtered paths, each of which falls incident onto one of the three sensors where each sensor acquires a single, complete band of the output image. The acquisition system then combines the three separate bands to form a single color image. The benefit of this technique is that sensor resolution is not sacrificed since the sensor resolution is identical to the resolution of the acquired image. These systems are more

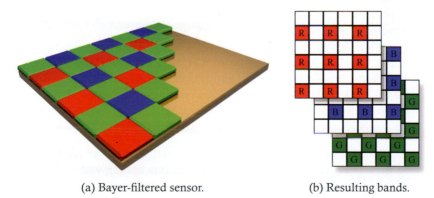

(a) Bayer-filtered sensor.   (b) Resulting bands.

Figure 3.13.   Demosaicing with the Bayer filter.

expensive, however, since more hardware is required and the optics must be precisely aligned in order to accurately register the three bands into a single composite output image.

Another new alternative involves a sophisticated image sensor developed by Foveon, Inc. Their sensor is based on photosites that are able to simultaneously detect the amount of red, green, and blue light. In this case, each sensor produces three separate light intensity readings and hence there is no need to sacrifice resolution for color depth. In addition, since there is a single image sensor, there is no need for multiple optical paths nor is there any need to register three separate images.

3.5   Display

Images are typically rendered on computer monitors, television screens, or printed media and each type of output device imposes constraints on how an image should be processed for display. Variations in the physical size and geometry of the display pixels and the total number of colors that an output is capable of displaying can greatly affect the quality of the displayed image.

As mentioned in the previous section, samples are dimensionless points implying that they cannot be viewed since they have neither width nor height. In order to make a digital image viewable, a reconstruction filter is applied that maps the samples of an image onto a display device. The reconstruction filter converts the collection of discrete samples into a continuous space which is amenable for display. Most cathode ray tube (CRT) monitors and most printers produce pixels that are circular in shape while other types of monitors produce rectangular pixels.

3.6   Exercises

1. Use Pixel Jelly to construct a checkerboard image (Edit $\rightarrow$ Create). The image should have a resolution of 250×250 and alternate between red and blue. Predict the result if the image is converted into one of RGB, CMY, YIQ, or HSB and a single band of the result is then displayed as a grayscale image. Justify your predictions and then use Pixel Jelly to check your prediction (Image Operations $\rightarrow$ Band Extract).

2. Equation (3.4) gives the transformation matrix for converting from the RGB to YIQ color space. The inverse of this matrix yields the transformation that converts from the YIQ color space to the RGB color space. Give the transformation matrix for converting from the YIQ to RGB color space and comment on the results.

3. Use Pixel Jelly to simulate Bayer filtering an image. Apply the Bayer filter operation to a color image of your choice and describe the visual result. Then convert the Bayer filtered image to grayscale and compare this to the grayscale version of the prefiltered color image. Explain the result.

4. Consider a 800×600 image that has been acquired by a system where each pixel has a 1.5:1 aspect ratio. Give the aspect ratio of the image.

5. Describe a demosaicing algorithm. Your procedure should accept a $M \times N$ 2D array of integers which are arranged in a Bayer filtered pattern and it should produce a $W \times H$ 2D array of colors. Indicate the relationship between M and W, N and H and be precise about how the colors are formed from the Bayer-filtered values. Discuss the strengths and weaknesses of your approach.

6. ⋆ The text claims on page 35 that the maximum L_1 distance between two points in HSB space is $1+2\sqrt{2}$ and that the maximum L_2 distance between two points in HSB space is $\sqrt{5}$. Give a proof that this claim is correct. Also give two colors C_1 and C_2 such that the L_1 distance between them is $1 + 2\sqrt{2}$ and the L_2 distance is $\sqrt{5}$.

Artwork

Figure 3.6. *Jeremiah* by Michelangelo di Lodovico Buonnicoti Simoni (1475–1564). Construction of the Sistine Chapel was completed in 1483 after nine years of construction and it is the best known place of worship in Vatican City. It took decades of further work to complete the interior decoration, which includes frescoes by some of the most talented and well-known Renaissance artists including Michelangelo, Raphael, and Botticelli. The decorative scheme of the chapel serves to depict, in visual form, the doctrine of the Catholic church from the creation narrative to the birth, life, and death of Christ, and then to the final day of judgment.

The image of Jeremiah was completed in approximately 1512 as part of a large fresco atop the ceiling. Jeremiah is one of the major prophets of Old Testament literature, who powerfully foretells of the destruction of Jerusalem due to the residents rebellion against God. Jeremiah is often known as the "weeping prophet" due to his great anguish over the people's refusal to repent and it is this posture of sorrow that is reflected here. Over time the chapel ceiling began to deteriorate and has since been restored. Although the image shown in this text is derived from the original fresco, it has been digitally enhanced for presentation.

Figure 3.8. *Napoleon at the Saint-Bernard Pass* by Jacques-Louis David (1748–1825). Jacques-Louis David was a well-known and influential French artist whose historical paintings were well suited to the political and moral climate of his time. David maintained a very strong following of students and his predominant style is known as academic Salon painting. He was an active proponent of the French Revolution and became something of the artistic censor of the French Republic. He was imprisoned after the fall of Robespierre but was eventually released with the regime change brought about by Napoleon I.

Napoleon commissioned David to commemorate his daring crossing of the Alps at St. Bernard Pass. This march had allowed the French to overcome the Austrian army and claim victory at the Battle of Marengo on June 14, 1800. Although Napoleon had crossed the Alps on a mule, he asked David to portray him mounting a fiery steed. The images shown in Figures 3.8 and 3.9 are enhanced reproductions of David's original, which was completed in 1801.

Figure 3.9. *Napoleon at the Saint-Bernard Pass* by Jacques-Louis David (1748–1825). See the description of Figure 3.8.

Figure 3.10. *The Gulf Stream* by Winslow Homer (1836–1910). Winslow Homer was born in Boston, Massachusetts, as the second of three sons and lived in Maine, close to the ocean, for most of his life. He first learned to paint from his mother, Henrietta, who was herself a skilled artist. Homer furthered his skills by working as a commercial illustrator, primarily for sheet music covers before finding later success and renown as an independent freelance artist.

Homer is best known for his vivid portraits of marine life involving ships, sailors, and the sea. The image of Figure 3.10 was inspired by trips made to the Bahamas in 1884–85 and 1898–99. The image is largely one of despair as it portrays a black sailor adrift and alone. The sailor apparently set out from Key West, Florida; the strong currents of the Gulf Stream push him into the Atlantic and a violent storm has snapped the mast and cut the rudder free from the ship. The boat is encircled with sharks where specks of red suggest that other sailors may have been lost overboard, and the threatening storm on the horizon insinuates that the worst is yet to come.

Figure 3.11. "Furman Owens" by Lewis Hine (1874–1940). Lewis Hine was an American photographer who used photography as a means of social reform by documenting and publicizing abuses within the child labor industry. His own words best summarize his view on child labor as it was practiced during his time: "There is work that profits children, and there is work that brings profit only to employers. The object of employing children is not to train them, but to get high profits from their work." Figure 3.11 is an image of Furman Owens. Hine's full title is "Furman Owens, 12 years old. Can't read. Doesn't know his A, B, C's. Said, 'Yes I want to learn but can't when I work all the time.' Been in the mills 4 years, 3 years in the Olympia Mill, Columbia, S.C."

Digital Images
in Java

4

4.1 Overview

The standard Java distribution includes support for loading images from files, writing images to files, viewing digital images as part of a graphical interface, and creating and editing digital images. This chapter provides an overview of commonly used image processing classes in Java and describes the fundamental concepts behind image data structures.

4.2 Image Structure

The central task of writing object-oriented image processing software is to generate data structures and methods flexible enough to represent images having different color and data models while still providing a uniform and efficient way of manipulating them. In order to understand Java's image processing class structure, we first examine the structure and essential nature of images in broad object-oriented terms. This section introduces a series of simple but illustrative image classes that are used to illuminate the design decisions behind Java's more sophisticated image processing library.

At the most basic design level, a digital image should be considered a three-dimensional structure having read and write support for both pixels and samples as well as methods for obtaining information about the dimensions of the image. Listing 4.1 shows a complete Java specification for the `DigitalImage` interface that supports these methods. This interface serves only to illustrate fundamental issues regarding Java's image processing library. This class is not used in Java's image processing library nor is it used throughout the remaining chapters of this text.

```
1  public interface DigitalImage {
2      public int getWidth();
3      public int getHeight();
4      public int getNumberOfBands();
5
6      public int[] getPixel(int x, int y);
7      public void setPixel(int x, int y, int[] pixel);
8      public int getSample(int x, int y, int band);
9      public void setSample(int x, int y, int band, int sample);
10 }
```

Listing 4.1. DigitalImage interface.

In Java, an interface is a collection of related methods without any bodies or code. An interface thus describes functionality without any restrictions on the way that the methods are implemented. Any class that implements an interface must provide the method bodies that are not given in the interface. Class implementation is similar to subclassing since classes that implement an interface are said to conform to the interface that they implement.

The DigitalImage interface defines seven methods that describe how clients are able to programmatically interact with and manipulate a digital image object. While the DigitalImage class places as few restrictions as practically possible on the way in which data is represented, the interface nonetheless makes a number of assumptions about how data is actually stored. For example, this interface defines a sample as an int while a pixel is represented as an array of samples.

While the interface requires information to flow between the client and supplier using particular data types, the interface does not make any assumptions about the way in which these operations are actually implemented. The next section highlights the various ways that this interface can be implemented and the performance implications of each technique.

4.2.1   Implementation Issues

In order to use the DigitalImage interface, implementing classes must be designed and implemented. It seems likely that most concrete classes will implement the get routines of the interface in precisely the same fashion. As a result, an abstract base class is presented in Listing 4.2, where the getWidth, getHeight, and getNumberOfBands methods of the AbstractDigitalImage class are completely implemented while leaving the others incomplete. This class is abstract as indicated by the class modifier, which simply means that the class is only partially implemented. The AbstractDigitalImage will serve as a base class for implementing classes that are themselves free to represent the image data and the sample processing functions in any way the designer chooses.

```
1  public abstract class AbstractDigitalImage implements DigitalImage {
2    protected int width, height, bands;
3
4    public AbstractDigitalImage(int w, int h, int b) {
5      width = w;
6      height = h;
7      bands = b;
8    }
9
10   public int getWidth() { return width; }
11   public int getHeight() { return height; }
12   public int getNumberOfBands() { return bands; }
13 }
```

Listing 4.2. `AbstractDigitalImage` class.

Listing 4.3 shows a complete implementation of the `DigitalImage` interface. Since the term *raster* is used in image processing to denote a rectangular grid of pixels, the `ArrayDigitalImage` class uses a variable named raster in which to store the pixel data. The raster variable is a three-dimensional array where the

```
1  public class ArrayDigitalImage extends AbstractDigitalImage {
2    private int[][][] raster;
3
4    public ArrayDigitalImage(int width, int height, int bands) {
5      super(width,height,bands);
6      raster = new int[height][width][bands];
7    }
8    public int getSample(int x, int y, int b) {
9      return raster[y][x][b];
10   }
11   public void setSample(int x, int y, int b, int s) {
12     raster[y][x][b] = s;
13   }
14   public int[] getPixel(int x, int y) {
15     int[] result = new int[bands];
16     System.arraycopy(raster[y][x], 0, result, 0, bands);
17     return result;
18   }
19   public void setPixel(int x, int y, int[] pixel) {
20     System.arraycopy(pixel, 0, raster[y][x], 0, bands);
21   }
22 }
```

Listing 4.3. The `ArrayDigitalImage` implementation of the `DigitalImage` interface.

first index corresponds to the row, the second index corresponds to the column, and the third index corresponds to the band. The `getSample` method must then take the indices given by the client and map them into the proper cell location of the raster. Note that since in Java arrays are indexed by row first and then by column, the `getSample` method switches the order of parameters x and y to match conventional array usage in Java. Also, since multidimensional arrays are implemented in Java as arrays of arrays, pixel-level access is exceptionally efficient and easily implemented as indicated by the brevity of the `getPixel` method.

The `System.arraycopy` method is a standard Java method that efficiently copies data from a source array, given as the first parameter, into a destination array, given as the third parameter. The index of the first source element is given as the second parameter while its corresponding location in the destination array is given as the fourth parameter. The number of elements to copy is given by the final parameter.

We will briefly analyze the performance of the `ArrayDigitalImage` class in terms of the speed of the individual methods and the amount of memory consumed by the object as a whole. Since there are a total of $W \cdot H \cdot B$ samples in a $W \times H \times B$ image we would expect that an efficient implementation would require at most $W \cdot H \cdot B$ bytes of memory. An analysis of the `ArrayDigitalImage` class shows that while there is efficient run-time support for the basic read and writes, the class consumes an excessive amount of memory. One source of unnecessary memory usage results from the fact that multidimensional arrays are implemented as arrays of arrays in Java.

Figure 4.1 illustrates the internal structure of the three-dimensional raster array for a $W \times H \times B$ digital image. Each arrow in the illustration is a reference (i.e., a pointer) to an array object and most likely, depending upon the specific platform, consumes 4 bytes of memory in the Java virtual machine

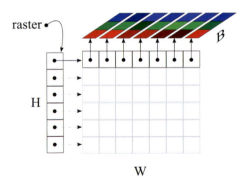

Figure 4.1. Structure of the three-dimensional array implementation of the `DigitalImage`.

(JVM). In the `ArrayDigitalImage` class, the raster array can be understood as a one-dimensional array of length H, where each array element is a reference to a one-dimensional array of length W, each element of which is a reference to a one-dimensional array of length B. In a $W \times H \times B$ image there are then $H + WH$ references that consume a total of $4(H + WH)$ bytes.

In addition, all array objects in Java contain a length attribute which is a 4-byte integer value. Since there are a total of $(1 + H + WH)$ arrays, an additional $4(1 + H + WH)$ bytes are consumed by these length attributes. Since the array references and length attributes are used only to support the internal structure of the implementation rather than directly representing image data, the memory consumed by these items is entirely redundant and can be considered as unnecessary overhead. The total amount of unnecessary memory consumed by this representation is given by

$$
\begin{aligned}
M_{\text{overhead}} &= 4(H + WH) + 4(1 + H + WH) \\
&= 8WH + 8H + 4 \\
&\simeq 8WH.
\end{aligned}
$$

The use of multidimensional arrays therefore consumes approximately 8 bytes per pixel in structural overhead alone and inflates the total memory requirements of a three-band image by a factor of approximately 2.5. Consider for example a 24 bpp RGB image that is 1000 pixels wide and tall. While we would expect that the image could be stored in approximately $24 \cdot 1000 \cdot 1000$ bits or 3,000,000 bytes, an `ArrayDigitalImage` object will consume well over $8 \cdot 1000 \cdot 1000$ or 8,000,000 bytes.

4.2.2   Linearization

The above shortcomings in terms of memory management can be overcome by forgoing the use of multidimensional arrays. We first change the type of the

Image Sample	Raster Index
$I(0,0)_0$	0
$I(0,0)_1$	1
$I(0,0)_2$	2
$I(1,0)_0$	3
$I(1,0)_1$	4
$I(1,0)_2$	5
$I(2,0)_0$	6
$\vdots$	$\vdots$
$I(x,y)_b$	$B(x + y \cdot W) + b$

Table 4.1. Raster layout of a $W \times H \times B$ digital image.

raster from a three-dimensional array to a single one-dimensional array. Since the raster must still contain all of the image samples, the length of the raster array must be equal to WHB. All image samples are then packed into the array by linearizing the indices in raster scan order. A *raster scan* is an ordered list of all pixels in an image starting at $I(0,0)$ and finishing at $I(W-1, H-1)_{B-1}$ proceeding in left-to-right, top-to-bottom fashion.

While the sample data is physically stored in a one-dimensional array the data actually *represents* a two-dimensional table. Since the samples of an image are accessed by their column, row, and band indices, we must develop a mapping between three indices, the column, row, and band indices, and the corresponding index in the one-dimensional raster array. Table 4.1 illustrates how the pixels of a digital image are packed into the raster. Given a $W \times H \times B$ image the raster element containing $I(x, y)_b$ is given by the last entry in Table 4.1. In other words, a request for sample $I(x, y)_b$ can be filled by returning the element at index $B(x + yW) + b$ of the raster.

```java
public class LinearArrayDigitalImage extends AbstractDigitalImage {
    private int[] raster;

    public LinearArrayDigitalImage(int width, int height, int bands) {
        super(width, height, bands);
        raster = new int[bands * width * height];
    }

    public int getSample(int x, int y, int b) {
        return raster[bands*(x+y*width)+b];
    }

    public void setSample(int x, int y, int b, int s) {
        raster[bands*(x+y*width)+b] = s;
    }

    public int[] getPixel(int x, int y) {
        int[] result = new int[bands];
        System.arraycopy(raster, bands*(x+y*width),
                         result, 0, bands);
        return result;
    }

    public void setPixel(int x, int y, int[] pixel) {
        System.arraycopy(pixel, 0,
                         raster, bands*(x+y*width), bands);
    }
}
```

Listing 4.4. Implementation of the `DigitalImage` interface using a one-dimensional array.

Using this technique, a linear array-based implementation of the `Digital Image` interface can be created, as shown in Listing 4.4, where the overhead of maintaining a multidimensional array has been eliminated by using a one-dimensional raster. Clients of the `LinearArrayDigitalImage` continue to access pixels by column and row even though the pixel values are stored in a one-dimensional raster. The run-time performance is comparable with that of the multidimensional implementation as little penalty is incurred by dynamically computing array indices from column, row, and band values.

4.2.3   Packing Samples into Pixels

While most image processing applications require a color depth of at most 24 bits per pixel, the `LinearArrayDigitalImage` still supports an excessive 96 bits per pixel since each pixel is composed of three 4-byte integers. Memory consumption can be further reduced by choosing a different implementation strategy for individual samples.

A more conservative choice is to implement each sample as a single `byte` rather than an `int`. While this solution may seem straightforward, a number of difficulties must be overcome. First of all, bytes are signed values in Java and represent values in the range -128 to $+127$, which doesn't correspond to the conventional image processing notion that an 8 bit sample should range in value between 0 (black) and 255 (white). In addition, the interface requires that `int`s be used externally, if not internally, as the native representation of a sample. Hence, some conversion must be performed for clients accessing samples through the `DigitalImage` interface.

Rather than making the raster a linear array of bytes, a more common technique is to pack an entire pixel into a single `int` value. Since an individual sample requires a single byte, it follows that a 4-byte integer has the capacity of holding four samples. Table 4.2 illustrates the idea behind packing four samples into a single `int` value. The table shows the standard Java RGB color representation, where the blue sample is encoded by bits 0 through 7, the green sample is encoded by bits 8 through 15, the red sample is encoded by bits 16 through 23 and the alpha (transparency) setting is encoded by bits 24 through 31.

A Java `int` is a single entity that is always interpreted as a signed integer. In this example, however, that value is not meaningful since the `int` is actually meant to represent four separate samples. Specialized functions must be

Byte 3 (Alpha)	Byte 2 (Red)	Byte 1 (Green)	Byte 0 (Blue)
11111111	01100001	00000111	11100001

Table 4.2. Example of a packed implementation. A single 4-byte integer value can store up to four 8-bit samples.

	Decimal	Binary
A	23523	0101101111100011
A >> 3	2940	0000101101111100
A << 5	31840	0111110001100000

Table 4.3. Illustration of Java's bit shifting operators.

written in order to access and interpret the individual samples of the pixel for reading or writing. Combining four samples into a single 32-bit int is known as *packing* while accessing a single sample is known as *unpacking*. Table 4.2 shows the packed RGB representation, which is Java's default technique for representing image data. Packing and unpacking rely on the use of Java's relatively obscure bit-level operators that allow the programmer to manipulate the individual bit patterns of an int. We will briefly review the bit-level operators in Java and then indicate how these operators are used to perform pixel packing and unpacking.

In the expression A>>B the right-shift operator, denoted as >>, shifts each bit of A to the right by B bits. The left-shift operator, as used in A<<B, shifts each bit of A to the left by B bits while zero filling from the least significant (rightmost) bit. The behavior of these operators is illustrated in Table 4.3, where the operand A is shown as a 16-bit value for the sake of presentation rather than its actual 32-bit value.

Java also supports bitwise *logical* operators (see Table 4.4), where the bit 1 is understood as *true* and the bit 0 is understood as *false*. In the expression A&B the bitwise logical and operator, denoted as & produces a bit that is the logical conjunction of its two operands. Bitwise logical or is denoted as | and bitwise logical exclusive or is denoted as ^. Logical negation is a unary operator denoted as ~, which flips all of the bits of its operand.

These operators can be used to pack and unpack pixels into a 32-byte int. Listing 4.5 gives methods that pack and unpack samples.

The getSampleFromPixel method unpacks a sample from a packed pixel. This method accepts a packed pixel and which band of the pixel (in the range [0,3]) to extract. This method first moves the desired band into the least significant 8 bits by right-shifting by either 0, 8, 16, or 24 bits. The method then returns the the bits 0 through 7 through bitwise conjunction. The 8-bit sample is then

	Decimal	Binary		Decimal	Binary
A	23523	0101101111100011	A	23523	0101101111100011
B	65280	1111111100000000	B	65280	1111111100000000
A & B	23296	0101101100000000	A | B	65507	1111111111100011

Table 4.4. An illustration of Java's bitwise logical operators.

```
1  int getSampleFromPixel(int pixel, int band) {
2      return (pixel >> band*8) & 0xFF;
3  }
4
5  int setSampleOfPixel(int pixel, int band, int sample) {
6      return ((0xFF&sample)<<band*8) | (pixel & ~(0xFF<<band*8));
7  }
```

Listing 4.5. The get and set methods for manipulating samples in packed pixels.

```
1  public class PackedPixelImage extends AbstractDigitalImage {
2      private int[] raster;
3      public PackedPixelImage(int width, int height, int bands) {
4          super(width, height, bands);
5          raster = new int[width * height];
6      }
7
8      public int getSample(int col, int row, int band) {
9          return (raster[col + row * width] >> (band * 8)) & 0xFF;
10     }
11
12     public void setSample(int col, int row, int band, int s) {
13         int pixelIndex = row * width + col;
14         int pixel = raster[pixelIndex];
15         int mask = 0xFF << (band * 8);
16         s = s << (band * 8);
17
18         raster[pixelIndex] = (pixel ^ ((pixel ^ s) & mask));
19     }
20
21     public int[] getPixel(int col, int row) {
22         int[] result = new int[bands];
23         for (int b = 0; b < bands; b++) {
24             result[b] = getSample(col, row, b);
25         }
26         return result;
27     }
28
29     public void setPixel(int col, int row, int[] pixel) {
30         for (int b = 0; b < bands; b++) {
31             setSample(col, row, b, pixel[b]);
32         }
33     }
34 }
```

Listing 4.6. `PackedPixelImage` implementation of `DigitalImage`.

returned in the least significant 8 bits of the resulting integer. Note that $0 \times FF$ represents the value 255 in hexadecimal notation, which is conventionally used when performing bitwise operations.

The `setSampleOfPixel` method accepts a pixel, the band of which should be modified, and the sample that should be written into the specified band. The method then copies the least significant 8 bits of the specified sample into the appropriate location of the pixel and returns the modified pixel value.

Using this pixel packing technique, a `PackedPixelArrayImage` class can be developed, which efficiently implements the DigitalImage interface. This class uses a one-dimensional array to store the pixel data and each pixel is represented by a packed `int`. The implementation is given in its entirety in Listing 4.6.

The `PackedPixelImage` provides the most compact memory footprint of the implemented images. Since, however, the `DigitalImage` interface defines a pixel as an array of sample values, the samples must be converted between bytes of a four-byte integer and elements of an array of integers when moving between external and internal representations. This conversion incurs a slight performance penalty when performing pixel-level read and writes due to the bit shifting overhead.

4.2.4   Indexing

While the packed array representation is memory efficient for any color image, an even more efficient representation can be used for a certain subset of color images. Consider a high-resolution image containing only a handful of colors: an image of the Olympic flag (containing only the colors blue, yellow, black, green, red, and white) or the image in Figure 4.2, which contains exactly 6 colors: black, white, gray, red, blue, and off-yellow. It seems extravagant to require 24 bits of memory per pixel for the image of Figure 4.2 since there are only six colors used in the entire image.

An indexed image uses a palette to assign each color in the image a unique number. The palette can be thought of as an array of colors where the number assigned to a color is its index in the array. An entire pixel can then be represented as the index of its color in the palette. Since the palette size is generally 256 or less, the result is to store entire pixels using 1 byte of memory rather than the 4 bytes required by the packed pixel representation. This technique is known as *indexing* since the pixel values are actually indices into an array of colors.

For example, consider that the image of Figure 4.2 uses an RGB palette consisting of only six colors. Since an RGB color consumes 3 bytes and there are six colors in the image, the palette itself will consume about 18 bytes of memory and have 0, 1, 2, 3, 4, 5 as the set of indices. Since each pixel of the image is an index into the palette, each pixel need only consume 3 bits of memory, which is smaller by a factor of eight over representations that require 24 bits per pixel.

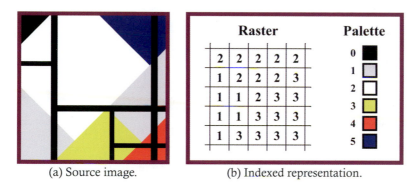

(a) Source image.   (b) Indexed representation.

Figure 4.2. Indexed image example.

The samples contained in the raster are only colors indirectly since the samples are actually indices into the palette. The raster shown in part (b) of Figure 4.2 depicts the samples that occur in the small subregion surrounding the intersection of the white, gray, and off-yellow elements in the source image.

It is important to note that modifying samples directly is not meaningful apart from the palette. Indexing also allows all samples of a particular color within a source image to be changed by simply replacing the old palette color with the new palette color. In order to change the blue region of Figure 4.2 to green, for example, we only need to replace the last color of the palette with green. None of the individual samples in the raster would need to be altered.

The amount of memory savings obtained by using indexing is dependent upon the number of colors in the image. Given a palette P, we denote the size of the palette by $|P|$. The number of bits per pixel is then given by $\lceil log_2(|P|) \rceil$. The GIF image file format, for example, uses this type of representation where the size of the palette is typically 256 and the size of each pixel is therefore 1 byte. In most cases the use of a palette will reduce the memory footprint to approximately $1/3$ of the memory footprint of the original image. Palettes do not typically exceed 256 colors since the memory savings in such cases are not substantial.

In addition to memory considerations, the indexing technique supports extremely efficient color modification. Changing all yellow pixels in the image to green, for example, is accomplished by making a single change in the palette where the color at index two is changed from yellow to green.

4.2.5 Summary

This section has presented three complete implementations of the `DigitalImage` interface and given an overview of an indexing implementation. Since each implementation conforms to the `DigitalImage` interface there is a uniform way of

	Multi-Dimensional Array	Linear Array	Packed Array	Indexed
Maximum bits per sample	32	32	8	unbounded
Maximum number of bands	unbounded	unbounded	4	unbounded
Memory footprint in bytes	$4WH(B+2)$	$4WHB$	$4WH$	$\frac{WHlog_2(|P|)}{8}$

Table 4.5. A comparison of the four `DigitalImage` implementations described in this section for a $W \times H \times B$ image.

manipulating each type of image, even though the internal representations vary in significant ways. Table 4.5 summarizes the features of each implementation. The multidimensional array implementation has no real advantage over the linear array and should not be used for any application. The linear array implementation supports 32 bits per sample and may be useful for applications that require more than eight bit depth. For most purposes, the packed array implementation is sufficient and makes a good choice. Color images rarely require more than 4 bands or 8 bits per pixel and for such images the packed array representation is concise and efficient. For the special class of images that contain few colors or where memory consumption must be minimized, indexing is appropriate.

4.3   Java's Imaging Library

The image processing library included with Java provides classes for representing images, processing images, and displaying images in a user interface. This library includes sophisticated classes that supports the kind of asynchronous image processing demanded by networked applications requiring responsive user interfaces. Since this text is not concerned with the development of sophisticated user interfaces, only the synchronous, also known as immediate mode image processing classes, will be covered.

The UML class diagram of Figure 4.3 depicts the immediate mode imaging classes covered in this text. The most important class is the `BufferedImage`. A `BufferedImage` object represents a digital image and is flexible enough to use different color models and different data representations similar those described in Section 4.2. Each `BufferedImage` contains a `WritableRaster` and a `ColorModel` object. A `WritableRaster` represents a rectangular array of pixels and a `ColorModel` defines the color space used by the image and a set of methods for translating between its native color space and the standard RGB color space.

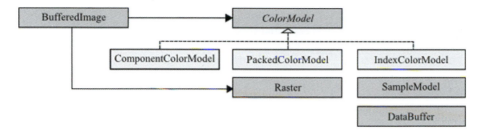

Figure 4.3. UML class diagram of the `BufferedImage` and associated classes.

Image processing operations are represented by the `BufferedImageOp` class, which is the standard class for single source image processing filters. Finally, the `ImageIO` class provides a few convenient methods for creating `BufferedImages` from image files and for writing `BufferedImages` to image files. Image processing applications will generally load a `BufferedImage` into memory through the `ImageIO` class, use a `BufferedImageOp` to process the image, and create an output image that is then saved to a file using the `ImageIO` class. These classes constitute the essential core of Java's image processing library and hence each of these classes is overviewed in the sections that follow.

4.3.1 BufferedImage

The `BufferedImage` class is similar to the `AbstractDigitalImage` presented in Section 4.2 since the primary ability of the class is to provide the height and width of the image in addition to getting or setting pixels in the image. Figure 4.4 lists the most important `BufferedImage` methods.

As expected, the `BufferedImage` getHeight and getWidth method provide the size of the image. The `getRGB` method returns the pixel at location (x, y), where the representation is an RGB packed-pixel integer regardless of the way the data is represented internally. In other words, the `getRGB` method may

java.awt.image.BufferedImage
`BufferedImage(int width, int height, int type)`
`int getHeight()`
`int getWidth()`
`int getRGB(int x, int y)`
`void setRGB(int x, int y, int rgb)`
`WritableRaster getRaster()`
`ColorModel getColorModel()`

Figure 4.4. `BufferedImage` methods (partial listing).

BufferedImage Constant	Description
TYPE_BYTE_BINARY	A packed 1, 2, or 4 bpp image
TYPE_BYTE_GRAY	An unsigned byte grayscale image
TYPE_BYTE_INDEXED	A byte valued indexed image
TYPE_INT_ARGB	An 8-bit packed sample RGB image with transparency
TYPE_INT_RGB	An 8-bit packed sample RGB image

Table 4.6. `BufferedImage` constants. Common choices of color model and raster.

convert the image data into standard form when providing the resulting pixel value. The `setRGB` method is symmetric to the `getRGB` method by overwriting the pixel at location (x, y) with the specified RGB value.

Since the `BufferedImage` class is extremely flexible in terms of its internal representation, constructing a `BufferedImage` can be a somewhat complex task. The width, height, number of bands, color model (RGB, grayscale, binary), color depth, and internal representation (indexed, packed, component) must be determined and provided to the constructor. Fortunately, the constructor shown in Figure 4.4 simplifies this task by giving predefined constants for the most common combinations of color model, color depth, and representation. This constructor requires only that the width and height of the image be given and a reference to a predefined constant that takes care of the remaining details.

Table 4.6 lists the predefined image types supported by the `BufferedImage`. The image types are simply static integer-valued constants of the `BufferedImage` class, where each constant begins with the prefix "TYPE" followed by a choice of representation (e.g., BYTE, INT) followed by the color model (e.g., BINARY, GRAY, INDEXED, ARGB). An image can then be programmatically constructed by supplying only three integer-valued parameters: the width, the height, and the static constant corresponding to the desired combination of color model and underlying type.

By way of example, the first line of Listing 4.7 constructs a 250-pixel wide by 300-pixel tall image using a grayscale color model where each sample is represented by a single byte. The second line constructs a 400×100 image using an RGB color model and packed sample representation. All `BufferedImages` constructed in this fashion are initially all black.

```
1   new BufferedImage(250, 300, BufferedImage.TYPE_BYTE_GRAY)
2   new BufferedImage(400, 100, BufferedImage.TYPE_INT_RGB)
```

Listing 4.7. Creating a black `BufferedImage`.

We now take note of what information a `BufferedImage` *cannot* provide. A `BufferedImage` cannot provide information about color depth or the number of bands, nor is it able to provide or process individual image samples. Even though the `BufferedImage` methods for reading and writing pixel data adopt the view that color is an 8 bit packed RGB, clients often require the ability to read and write individual samples of a pixel using color models and bit depths other than packed RGB. While sample-level processing is not directly supported by `BufferedImages`, the samples of an image can be processed by use of the images `ColorModel` and `WritableRaster`, which are obtained by the `getColorModel` and `getWritableRaster` methods.

4.3.2   ColorModel

The primary task of the `ColorModel` is to analyze a pixel and extract the individual color components: red, green, and blue, for example. `ColorModel` is an abstract class with subclasses that support packed pixel data (`PackedColorModel`), indexed pixel data (`IndexColorModel`), and direct representation of samples (`ComponentColorModel`). Since the RGB color space is the most common in general computing systems, RGB serves as a default and all `ColorModels` are able to convert from their native format to packed `int` RGB format.

4.3.3   WritableRaster

The `WritableRaster` represents, at a conceptual level, a two-dimensional array of pixels. Table 4.7 lists the most commonly used raster methods. Since a raster is a two dimensional table of pixels, it supports accessor methods to obtain the width and height of the table in addition to the number of bands in the image. Although clients are able to read and write pixels through the `BufferedImage` getRGB and setRGB methods, reading and writing samples must be done through the getSample and setSample methods of the `WritableRaster` class. These methods make no assumption about the color space nor the sample representation and hence provide a more general interface to pixel data than the methods provided by the `BufferedImage`.

java.awt.image.WritableRaster
int getWidth()
int getHeight()
int getNumBands()
int getSample(int x, int y, int b)
void setSample(int x, int y, int b, int sample)

Table 4.7. Raster methods (partial listing).

4.3.4 ImageIO

The most convenient way of creating a `BufferedImage` is by loading data from an image file. The `ImageIO` class, defined in the `javax.image.io` package, contains a collection of convenient methods to read `BufferedImages` from a file and to save them into a file. Table 4.8 shows the two primary methods of the `ImageIO` class.

javax.imageio.ImageIO
`BufferedImage read(File f)`
`BufferedImage write(BufferedImage im, String format, File file)`

Table 4.8. ImageIO read and write methods.

The `read` method accepts a `File` object that must be an image file in any of the BMP, GIF, PNG, or JPEG file formats. The `read` method will open the file and construct a `BufferedImage` object corresponding to the information in the file. It is important to understand that once the file is read and a `BufferedImage` constructed, there is no longer any connection between the `BufferedImage` and file. While the `BufferedImage` that is created by the `read` method is an accurate depiction of the image file, there is no correlation between the representation of the `BufferedImage` and the file itself. The file, for example, may use an indexed color model given in YCbCr color space while the `BufferedImage` may use a direct color model given in RGB space.

The `write` method is symmetric with the `read`. This method takes a `BufferedImage`, a string that identifies the file format to use for encoding, and a file in which to store the encoded image. The `ImageIO` class provides an extremely useful and simple method for reading images from disk and can also read images directly from networks through URLs. The code fragment shown in Listing 4.8 shows how to use the `ImageIO` class to convert an image from PNG to GIF format.

```
1 BufferedImage image=ImageIO.read(new File("Filename.png"));
2 ImageIO.write(image, "GIF", new File("Filename.gif"));
```

Listing 4.8. Loading and saving a BufferedImage with ImageIO.

4.3.5 BufferedImageOp

Once an image is constructed by either reading data from an image file or calling a constructor, image processing methods typically take a `BufferedImage` object

```
1  BufferedImage source = ImageIO.read(new File("filename.png"));
2  BufferedImage destination = new BufferedImage(source.getWidth(),
3                                                 source.getHeight(),
4                                                 source.getType());
5
6  WritableRaster srcRaster = source.getRaster();
7  WritableRaster destRaster = destination.getRaster();
8  for(int row=0; row<srcRaster.getHeight(); row++) {
9    for(int col=0; col<srcRaster.getWidth(); col++){
10     for(int band=0; band<srcRaster.getNumBands(); band++) {
11       int sample = srcRaster.getSample(col,row,band);
12       destRaster.setSample(col,row,band,255-sample);
13     }
14   }
15 }
16
17 ImageIO.write(destination, "JPEG", new File("output.jpg"));
```

Listing 4.9. Loading and inverting an image.

and construct a new image by applying some some pixel processing algorithm. A single source image filter is one that accepts a single image as input and produces a singe image as output. The input image is known as the source and the output image is known as the destination.

The code fragment of Listing 4.9 loads a PNG image and creates its negative. The negative of an image is one where all bright pixels become correspondingly dark and all dark pixels become correspondingly bright. In formal terms, the inverse of a sample S is defined as $S_{\text{inverse}} = 255 - S$ for an 8-bit sample. In Listing 4.9 the source image is read from a file and the destination image is programmatically constructed such that it has the same width, height, and color model as the source image. Since the destination image after construction is fully black, the destination samples are then set to their proper values by iterating over every sample of the source and writing their inverse into the corresponding destination locations.

The code fragment in Listing 4.9 illustrates the basic pattern of the vast majority of single-source image processing algorithms. Since single-source algorithms represent a such a large and important class of image processing operations, Java uses the `BufferedImageOp` interface to formalize and standardize this process. Table 4.9 gives the methods of the `BufferedImageOp` interface.

The central method is `filter`, which performs a single source image processing operation whereas all of the other methods serve supportive roles. The source and destination images are passed as input parameters to the `filter` method and

java.awt.image.BufferedImageOp

`BufferedImage createCompatibleDestImage(BufferedImage src, ColorModel destCM)`
`BufferedImage filter(BufferedImage src, BufferedImage dest)`
`Rectangle2D getBounds(BufferedImage src)`
`Point2D getPoint2D(Point2D srcPt, Point2D destPt)`
`RenderingHints getRenderingHints()`

Table 4.9. Methods of the `BufferedImageOp` interface.

the destination image is then returned. If the destination image is not provided (i.e., if it is null) the filter method constructs a `BufferedImage` object, which is designated as the destination, fills in its samples appropriately, and returns it as a result. Construction of the destination image is the responsibility of the `createCompatibleDestImage` method.

It is important to note that the filter method *should never modify the source image* since it is implicitly understood as a read-only input parameter while the destination is a read-write output parameter. While this design principle is always followed by `BufferedImageOp` implementations, it can be skirted by clients that call the filter method and pass in the same object as both the source and destination. Under such a scenario the source image is modified by the filter since the source image is also the destination. Many filters will not function correctly

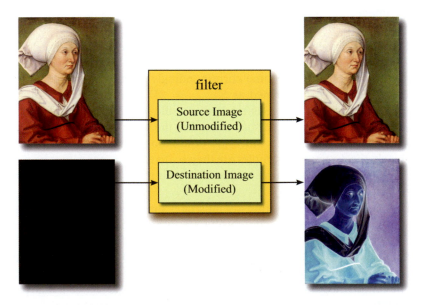

Figure 4.5. Illustration of the `BufferedImageOp`.

if the client passes in the same image object as both the source and destination, but the class of filters that does function properly under such a scenario is known as *in place* filters. Clients must be careful to avoid using the same object as both source and destination unless the filter is described as an in place operation.

Figure 4.5 illustrates the BufferedImageOp filtering process for a filter that negates the source. Prior to the filtering operation, the destination image is black and possesses the same dimensions and color space as the source. The filter method scans the source image without modifying it and updates the pixels of the destination to create an image that is the negative of the source.

The getPoint2D, getRenderingHints, and getBounds methods will be discussed later in the text and are largely unused in this text.

4.3.6   Custom Filters with the Imaging Package

Throughout this text, we develop a set of customized image processing classes that are well integrated with Java's standard image processing library. These classes are meant to simplify the development of image processing applications while maintaining solid software engineering principles. The classes are contained in the imaging package, which has subpackages named features, gui, utilities, ops, io, and scanners. Figure 4.6 shows the package structure of this customized code presented in this text. In this section we describe a handful of the core classes in the imaging package.

First we note that iterating through each sample of an image is an exceptionally common task that can be accomplished by nested loops as was illustrated in Listing 4.9. The process of visiting each sample of an image is known as *scanning*. Listing 4.10 shows how each sample of an image can be processed by traversing

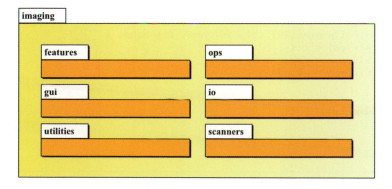

Figure 4.6. The imaging package structure.

```
1  for(int row=0; row<srcRaster.getHeight(); row++) {
2    for(int col=0; col<srcRaster.getWidth(); col++){
3      for(int band=0; band<srcRaster.getNumBands(); band++) {
4        int sample = srcRaster.getSample(col, row, band);
5        // process the sample ...
6      }
7    }
8  }
```

Listing 4.10. Pattern for iterating over every image sample.

the samples in top-to-bottom, left-to-right fashion using nested loops. The structure of this code fragment lends itself to two common errors that are addressed by the Location and Scanner classes described in the following sections.

Location. One of the most common mistakes made when developing customized image processing filters is to incorrectly scan the image by mistaking the row index for the column or the column index for the row. I have spent many hours myself debugging otherwise well-written code that inadvertently swapped the row with the column indices. In order to minimize this common error and to shorten the amount of code needed to iterate over image samples, we develop two collaborating classes, Location and RasterScanner, that allow clients to more easily traverse the samples of an image. Both of these classes are found within the pixeljelly.scanners package.

```
1  public class Location {
2    public int row, col, band;
3
4    public Location(int col, int row, int band) {
5      this.col = col;
6      this.row = row;
7      this.band = band;
8    }
9
10   public Location() {
11     this(0,0,0);
12   }
13 }
```

Listing 4.11. Partial listing of the Location.

The Location is a very simple class representing a single object containing the three separate indices needed to access an image sample. In other words, a single Location object can be used to identify a sample rather than using three separate index values. The Location class is partially shown in Listing 4.11. Note that the attributes row, col, and band are public and hence directly accessible by any client.

RasterScanner. The second class, RasterScanner, is a class that serves as an iterator by producing Location objects one after the other. In Java, an iterator is used to explicitly traverse the elements of a sequence and in this case the elements are the indices of image samples. Since iterators underlie the ability of loops to traverse the elements of a Collection, we can use the RasterScanner in the context of Java's specialized for-each loops. While the implementation of RasterScanner is not listed due to its rather sophisticated class structure and reliance on generics, it is straightforward to use. A full description of this class is included in Appendix B.

A RasterScanner object is constructed by supplying a BufferedImage and a Boolean flag indicating whether the samples of the image or the pixels are to be visited. If individual samples are to be iterated then the flag is set to true; otherwise, when set to false, the band is always zero while the column and row indices scan over the image. By using these two classes together, we can now simplify the code of Listing 4.10 to that of Listing 4.12. Listing 4.12 should be understood as meaning *for every sample s in image src, process s and place it in the destination*.

```
1  for(Location pt : new RasterScanner(src, true)) {
2      int s = srcRaster.getSample(pt.col, pt.row, pt.band);
3      // process the sample ...
4      dst.getRaster().setSample(pt.col, pt.row, pt.band, s);
5  }
```

Listing 4.12. Improved pattern for iterating over every image sample.

NullOp. Since image processing operations in Java are implementations of the BufferedImageOp interface, we must provide a concrete implementation for each custom filter that we write. Towards this end we introduce a general base class implementation that can be easily subclassed to provide specific image processing functions. The NullOp is a custom filter defined in the pixeljelly.ops package that simply copies each source sample to the destination (see Listing 4.13). The NullOp class is a complete implementation of Buffered ImageOp and is meant to be subclassed by overriding the filter method.

```
1  public class NullOp implements BufferedImageOp {
2
3    public BufferedImage createCompatibleDestImage(BufferedImage src,
4                                                    ColorModel destCM) {
5      return new BufferedImage(
6        destCM,
7        destCM.createCompatibleWritableRaster(src.getWidth(), src.getHeight()),
8        destCM.isAlphaPremultiplied,
9        null);
10     }
11
12   public BufferedImage filter(BufferedImage src, BufferedImage dest) {
13     if (dest == null) {
14       dest = createCompatibleDestImage(src, src.getColorModel());
15     }
16
17     WritableRaster srcRaster = src.getRaster();
18     WritableRaster destRaster = dest.getRaster();
19     for (Location pt : new RasterScanner(src, true)) {
20       int sample = srcRaster.getSample(pt.col, pt.row, pt.band);
21       destRaster.setSample(pt.col, pt.row, pt.band, sample);
22     }
23
24     return dest;
25   }
26
27   public Rectangle2D getBounds2D(BufferedImage src) {
28     return src.getRaster().getBounds();
29   }
30
31   public Point2D getPoint2D(Point2D srcPt, Point2D dstPt) {
32     if (dstPt == null) {
33       dstPt = (Point2D) srcPt.clone();
34     } else {
35       dstPt.setLocation(srcPt);
36     }
37
38     return dstPt;
39   }
40
41   public RenderingHints getRenderingHints() {
42    return null;
43   }
44 }
```

Listing 4.13. NullOp.

As mentioned earlier, the createCompatibleDestImage method is responsible for programmatically constructing a destination image whenever the destination given to the filter method is null. In our implementation, the destination will be structurally identical to the source image since it will use the same color model, representation, and dimension. For many types of filters, however, the destination image will be of a different kind than the source and hence this method must occasionally be overridden in subclasses.

The getRenderingHints method can be trivial to implement since the method is intended only to allow systems to optimize rendering performance for specific platforms when possible. Our implementation provides no hints for optimized rendering. The getBounds2D and getPoint2D methods do not have any real significance for this example (nor indeed for many types of filtering operations) and are also not discussed further in this text. These methods may be safely ignored throughout the remainder of this text.

InvertOp. The InvertOp class of Listing 4.14 rewrites the code of Listing 4.9 using the BufferedImageOp and scanner classes. While the earlier code fragment for image negation was correct, it was not integrated into Java's image processing library but was an isolated method. Note how the earlier code fragment is now placed into the filter method of a BufferedImageOp class. When used in this form, the source image must be created by the client and then passed in as a parameter while the destination *may* be created by the client; otherwise it will be created via the createCompatibleDestImage of the BufferedImageOp.

Listing 4.15 finally gives a code fragment showing how a client may use the InvertOp class to filter a BufferedImage. In this example, a BufferedImage is created by loading data from a PNG file while the destination is not provided

```
1  class InvertOp extends NullOp {
2    public BufferedImage filter(BufferedImage source, BufferedImage dest) {
3      if(dest == null) {
4        dest = createCompatibleDestImage(src, src.getColorModel());
5      }
6
7      WritableRaster srcRaster = source.getRaster();
8      WritableRaster destRaster = dest.getRaster();
9      for (Location pt : new RasterScanner(src, true)) {
10       int sample = srcRaster.getSample(pt.col, pt.row, pt.band);
11       destRaster.setSample(col, row, band, 255 - sample);
12     }
13     return dest;
14   }
15 }
```

Listing 4.14. InvertOp.

```
1  BufferedImage src=ImageIO.read(new File("Filename.png"));
2  BufferedImageOp op = new InvertOp();
3  BufferedImage result = op.filter(src, null);
4  ImageIO.write(result, "PNG", new File("Inverted.png"));
```

Listing 4.15. Applying the `NegativeOp` to a source image.

and hence is nulled. The filter method of the `InvertOp` is then invoked, passing in references to the source and destination images. Since the destination value is null, the filter method constructs an image of the proper size and color model and inverts each sample of the source into the destination, which is then returned as the result.

As with all interfaces, the `BufferedImageOp` describes the expected behavior of its methods while providing no information about how these methods are actually implemented. Concrete implementations provide specific code appropriate for the specific operation that they perform. `AffineTransformOp`, `ColorConvertOp`, `ConvolveOp`, `LookupOp`, and `RescaleOp` classes are `BufferedImageOp` implementations bundled with Java's image processing library. Programmers may also create their own customized filters by implementing the interface as we have done with the `InvertOp` class.

4.3.7 Image Display Classes

Most image processing applications must, of course, be able to display images. Java provides a collection of classes that are useful for constructing graphical interfaces and supplying a modicum of support for displaying images. Since the intent of this text is to explain fundamental image processing techniques, coverage of these classes will only extend as far as necessary for constructing the simplest type of graphical applications.

Graphics Methods and Description
`void drawLine(int x1, int y1, int x2, int y2)`   draws a line between points (x1,y1) and (x2,y2)
`void drawOval(int x, int y, int width, int height)`   draws an oval at location (x,y) of the specified width and height
`void drawRect(int x, int y, int width, int height)`   draws a rectangle at location (x,y) of the specified width and height
`void drawImage(Image im, int x, int y, ImageObserver obs)`   the image is drawn with the upper-left corner at location (x,y)

Table 4.10. The most commonly used methods of the `Graphics` class.

```
1  public class ImageComponent extends JComponent {
2    protected BufferedImage image;
3
4    public BufferedImage getImage() {
5      return image;
6    }
7
8    public ImageComponent(BufferedImage image) {
9      setImage(image);
10   }
11
12   public void setImage(BufferedImage image) {
13     this.image = image;
14     Dimension dim;
15     if(image != null ){
16       dim =new Dimension(image.getWidth(), image.getHeight());
17     } else {
18       dim = new Dimension(50,50);
19     }
20
21     setPreferredSize(dim);
22     setMinimumSize(dim);
23     revalidate();
24     repaint();
25   }
26
27   public void paintComponent(Graphics gin) {
28     gin.setColor(getBackground());
29     gin.fillRect(0, 0, getWidth(), getHeight());
30     if(image != null) {
31       Insets ins = getInsets();
32       int x = (getWidth() - image.getWidth())/2 - ins.left;
33       int y = (getHeight() - image.getHeight()) / 2 - ins.top;
34       gin.drawImage(image, x, y, this);
35     }
36   }
37
38   public static showInFrame(String title , BufferedImage src) {
39     JFrame win = new JFrame(title);
40     win.add(new JScrollPane(new ImageComponent(src)), BorderLayout.CENTER);
41     win.pack();
42     win.setVisible(true);
43   }
44 }
```

Listing 4.16. A class for displaying a `BufferedImage`.

Java provides a suite of classes that are designed for developing interactive graphical applications. This suite, known collectively as the Swing library, contains graphical components such as text fields, buttons, panels, borders, and the like. The base class for all Swing components is the JComponent that represents, in an abstract sense, a rectangular region of the output device. The location of a JComponent is given as the Cartesian coordinate of its upper-left-hand corner while the size is given as integer-valued width and height. In addition to these geometric properties, the JComponent is responsible for painting itself onto the computer screen when requested.

Whenever the Swing framework determines that a component must be drawn onto the computer screen (as a result of resizing a window or uncovering a window, for example), the framework calls the JComponent's paintComponent method. This method is given a Graphics object which can be used to draw lines, ovals, and other graphical elements onto the output device. The Graphics object is an abstraction that may represent a computer screen, a printer, or any other type of output device. A JComponent need only utilize the drawing commands supplied by the Graphics object in order to draw itself onto the output device. Table 4.10 summarizes the most commonly used drawing commands of the Graphics class.

Customized components can be integrated into the Swing library by subclassing existing JComponents and providing, at a minimum, a paintComponent method. The ImageComponent of Listing 4.16 gives a complete implementation of a customized Swing component that directly subclasses JComponent.

As required, the class provides a paintComponent method that simply commands the Graphics object to draw an image onto the component. Listing 4.17 gives a code fragment that reads an image from a file, copies the image using the NullOp, and displays both the source and destination in separate windows.

```
1 BufferedImage src = ImageIO.read(new File("annunciation.jpg"));
2 ImageComponent.showInFrame("source", src);
3 BufferedImage dest = new NullOp().filter(src, null);
4 ImageComponent.showInFrame("example", dest);
```

Listing 4.17. Loading and displaying an image in a window.

4.3.8 Pixel Jelly

This text is supported by an image processing application named Pixel Jelly. The application is written using many of the classes that are developed throughout this text and is also designed to be extensible so that external filters can be used for experimentation. Any class that implements both the BufferedImageOp and PluggableImageOp interfaces can be loaded into the Pixel Jelly application and used as though it were a native part of the software. The PluggableImageOp

PluggableImageOp
`BufferedImageOp getDefault(BufferedImage src)` `String getAuthorName()`

Table 4.11. `PluggableImageOp` interface.

interface is defined in the `pixeljelly.ops` package, as shown in the UML class diagram of Table 4.11.

The `getAuthorName` method allows Pixel Jelly to display the name of the plugin author. The body of this method will typically be a single line of code returning your name. The `getDefault` method is responsible for constructing a default instance of the `BufferedImageOp` for use on the provided source image. The Pixel Jelly application uses this method to obtain an instance of your `BufferedImageOp` whenever a plugin is used. Listing 4.18 shows how the `InvertOp` class can be made into a plugin by implementing the pluggable interface and providing the authors name and the default operation for the `InvertOp`.

```
1  class InvertOp extends NullOp implements PluggableImageOp {
2    public BufferedImage filter(BufferedImage source, BufferedImage dest) {
3      // code not shown
4    }
5
6    public String getAuthorName() {
7      return "Kenny Hunt";
8    }
9
10   public BufferedImageOp getDefault(BufferedImage src) {
11     return new InvertOp();
12   }
13 }
```

Listing 4.18. `InvertOp` as a `PluggableImageOp`.

4.4 Exercises

1. Implement the `DigitalImage` interface of Listing 4.1 using a palette-based indexing technique. Your class should conform to the UML class diagram listed below and should not consume more than 8 bits per pixel.

≪ *implements*   *DigitalImage* ≫
IndexedDigitalImage
+IndexedDigitalImage(width:int, height:int)
+IndexedDigitalImage(width:int, int height:int, palette:Color[])
+setPaletteColor(paletteIndex:int, color:Color) : void
+getPaletteColor(paletteIndex:int) : Color

2. Complete the `DigitalImageIO` class that is outlined below. The central purpose of this class is to allow `DigitalImages` to be loaded from and saved as ASCII-encoded PPM. The kind of digital image constructed by the load method should be either an `IndexedDigitalImage` (see problem 1 above), `PackedPixelImage`, `LinearArrayDigitalImage` or an `ArrayDigitalImage` depending upon the value of the `imageType` parameter supplied by the client. A specification of the PPM file format can be found at http://netpbm.sourceforge.net/doc/ppm.html. If the `imageType` is `INDEXED` then the PPM file must consist of at most 256 unique colors in order for the load method to succeed.

```
1  import java.io.*;
2
3  public class DigitalImageIO {
4    enumeration ImageType {INDEXED, PACKED, LINEAR_ARRAY, MULTI_DIM_ARRAY};
5    public static DigitalImage read(File file, ImageType type)
6    throws IOException {...}
7    public static void write(File file, DigitalImage image)
8    throws IOException {...}
9  }
```

3. Write a class that contains two static methods. The first method is named `toBufferedImage`. This method accepts a `DigitalImage` object and produces an equivalent `BufferedImage`. The second method is named `toDigitalImage`. This method accepts a `BufferedImage` object and produces an equivalent `DigitalImage` object. An outline of the class is provided below.

```
1  import java.awt.image.*;
2
3  public class ImageConvertor {
4      public static BufferedImage toBufferedImage(DigitalImage src) {...}
5      public static DigitalImage toDigitalImage(BufferedImage src) {...}
6  }
```

4. Implement the `BufferedImageOp` interface by authoring a `TopBottom FlipOp`. The filter method should support in place performance and simply flip the source image about the horizontal axis.

5. Implement the `BufferedImageOp` interface by authoring a `LeftRight FlipOp`. The filter method should support in place performance and simply flip the source image about the vertical axis.

6. Implement the `BufferedImageOp` interface by authoring a `RotateOp`. The filter method should support in place performance and rotate an image by either 90° clockwise, 90° counterclockwise, or by 180°. Note that the destination image may have different dimensions than the source and hence the `createCompatibleDestImage` method must be overridden in the `RotateOp`.

7. Implement the `BufferedImageOp` interface by authoring a `Transpose FlipOp`. The filter transpose the rows and columns of the image. Note that the destination image will have different dimensions than the source and hence the `createCompatibleDestImage` method must be overridden in the `DiagonalFlipOp`.

8. Implement the `BufferedImageOp` interface by authoring a `CenterFlipOp`. The filter method should support in-place performance and reflect the source image about the *center point* of the source image. HINT: You may wish to make use of the `HorizontalFlipOp` and the `VerticalFlipOp` from problems 4 and 5.

9. Write a program that has a graphical user interface. The program will allow users to select an image file in PNG, GIF, or JPEG format and then view the image on the computer screen. In addition, the program should contain a way to display the properties of the image; including width, height, number of bands, and the type of `SampleModel`.

Artwork

Figure 4.2. *Simultaneous Counter-Composition* by Theo van Doesburg (1883–1931). Theo van Doesburg was a Dutch painter who explored a variety of genres during his career but is most closely associated with the De Stiljl art movement of which he was the founder. He collaborated closely with Piet Mondrian to advance non-representational geometric abstractions created of strong vertical and horizontal lines and primary colors. They were close collaborators until about 1924 at which time a debate over the legitimacy of diagonal lines caused a rift between the two. The image of Figure 4.2 is a simplified rendition of his original work and is illustrative of the fact that van Doesburg argued for the the use of diagonal lines while Mondrian renounced them altogether.

Figure 4.5. *Portrait of Barbara Dürer, born Holper* by Albrecht Dürer (1471–1528). Albrecht Dürer was a German artist proficient in engraving, painting, and print-making. He was born in 1471 and died in 1528 at the age of 56. His work was strongly rooted in mathematical theories of perspective and proportion. Figure 4.5 is a portrait of his mother and was completed around 1491. Barbara gave birth to 18 children over her 25 years of marriage, but only three of these survived beyond infancy. Amid this tragic situation, she was nonetheless blessed to give birth to one of the greatest Renaissance artists in Northern Europe.

Point Processing Techniques 5

5.1 Overview

The goal of image enhancement is to improve the quality of an image with respect to a particular application. Image enhancement, for example, may be used to brighten an underexposed digital photo for viewing in a photo album or to stretch the contrast of an image to compensate for defective acquisition hardware. Since the objective of image enhancement can only be defined with respect to a specific application, this chapter presents the underlying techniques of image enhancement and provides examples that illustrate how these techniques can be applied to solve particular image enhancement problems.

Two fundamentally distinct types of image enhancement techniques exist: those that operate within the spatial domain and those that operate in the frequency domain. While these terms are more clearly delineated in Chapter 9 we informally define the spatial domain as a way of representing an image by spatially arranging pixels (colors) to form a complete whole. The frequency domain is an alternate way of representing visual information by describing patterns of global fluctuation in images. This chapter focuses exclusively on spatial domain techniques, leaving the more difficult frequency domain issues for Chapter 9.

Spatial domain techniques can be further classified as either point or regional processing. Under point processing a single input sample is processed to produce a single output sample even though many point processing operations allow for limited regional pre-processing. In regional processing the value of the output sample is dependent on the values of samples within close proximity to the input sample. Exactly how the region is selected and how samples within the region affect the output depends upon the desired effect. Chapters 5 through 8 focus exclusively on spatial domain techniques, leaving the more difficult frequency domain issues for Chapter 9.

To summarize then, we see that image enhancement techniques are done in either the frequency or spatial domain. Within the spatial domain we note that a technique can be classified as either a point processing or regional technique. Figure 5.1 shows this classification scheme.

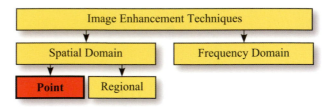

Figure 5.1. Taxonomy of image processing techniques.

5.2 Rescaling (Contrast and Brightness)

Rescaling is a point processing technique that alters the contrast and/or brightness of an image. In digital imaging, exposure is a measure of how much light is projected onto the imaging sensor. An image is said to be *overexposed* if details of the image are lost because more light is projected onto the sensor than what the sensor can measure. An image is said to be *underexposed* if details of the image are lost because the sensor is unable to detect the amount of projected light. Images which are underexposed or overexposed can frequently be improved by brightening or darkening them. In addition, the overall contrast of an image can be altered to improve the aesthetic appeal or to bring out the internal structure of the image. Even nonvisual scientific data can be converted into an image through adjusting the contrast and brightness of the data.

5.2.1 Definition

Given an input sample S_{input} a linear scaling function can be used to generate an output sample value S_{output}. The value of the output sample is determined by linearly scaling the input, as shown in Equation (5.1):

$$S_{\text{output}} = \alpha \cdot S_{\text{input}} + \beta. \tag{5.1}$$

In this equation, α is a real-valued scaling factor known as the gain and β is a real-valued offset known as the bias. While Equation (5.1) shows how to rescale a single sample, it is apparent that an entire image is rescaled by applying this linear scaling function to every sample in the source image.

To determine the effect of bias on the rescaling function, let's assume that $\alpha = 1$. When the bias is greater than 0 the output image is brightened by the amount β since every sample increases by the same amount. When the bias is less than 0 the output image will be darkened by the amount of offset, and if the offset is 0 then the output image is a copy of the input.

To determine the effect of the gain on the rescaling function, let's assume that $\beta = 0$. When the gain is less than 1 the image is darkened since all output sample values are smaller than the input samples. When the gain exceeds 1 the image is brightened since all output sample values are increased and when the gain is equal to 1 the image is unaltered. More importantly, however, the gain affects the contrast of the image. The term *contrast* refers to the amount of difference between various regions of an image. High-contrast images possess a larger variation in luminosity than low contrast images. In high-contrast images, different-colored regions are clearly distinguishable while in low-contrast images, different-colored regions are more difficult to identify.

The effect of gain and bias on contrast can be seen by considering two samples of an input image which we will refer to as S_1 and S_2. The rescaled samples are obtained from Equation (5.1):

$$\begin{aligned} S_1' &= \alpha S_1 + \beta, \\ S_2' &= \alpha S_2 + \beta. \end{aligned} \qquad (5.2)$$

Since contrast deals with the difference between two samples, we here define contrast as the magnitude of the difference between them. The magnitude of the difference between the two input samples is denoted as $\triangle S$ and the magnitude of the difference between the two rescaled samples is denoted as $\triangle S'$. The differences are given by Equation (5.3):

$$\begin{aligned} \triangle S' &= |S_1' - S_2'|, \\ \triangle S &= |S_1 - S_2|. \end{aligned} \qquad (5.3)$$

The relative change in contrast between the two samples in the source image and the corresponding rescaled samples is then given by Equation (5.4):

$$\frac{\triangle S'}{\triangle S} = \frac{|S_1' - S_2'|}{|S_1 - S_2|}. \qquad (5.4)$$

Substitution of Equation (5.2) into Equation (5.4) yields

$$\begin{aligned} \frac{\triangle S'}{\triangle S} &= \frac{|(\alpha S_1 + \beta) - (\alpha S_2 + \beta)|}{|S_1 - S_2|} \\ &= \frac{|\alpha| \cdot |S_1 - S_2|}{|S_1 - S_2|} \\ &= |\alpha|. \end{aligned}$$

From this we see that gain controls the overall contrast of an image while the bias has no effect on the contrast. The bias, however, controls the overall intensity or brightness of the rescaled image. When the bias is a positive value, rescaling serves to brighten the output image; when the bias is negative the image is darkened, and a zero bias will have no effect on the intensity of the output.

S	$\beta = 0$			$\beta = 50$		
	$\alpha = .25$	$\alpha = 1$	$\alpha = 1.5$	$\alpha = .25$	$\alpha = 1$	$\alpha = 1.5$
0	0	0	0	50	50	50
1	0	1	1	50	51	51
2	0	2	3	50	52	53
3	0	3	4	50	53	54
4	1	4	6	51	54	56
$\vdots$	$\vdots$	$\vdots$	$\vdots$	$\vdots$	$\vdots$	$\vdots$
253	63	253	379	113	303	429
254	63	254	381	113	304	431
255	63	255	382	113	305	432

Table 5.1. The effect of various gain and offsets as applied to 8-bit samples.

The effect of gain and bias is shown numerically in Table 5.1. For the input samples given in the column labeled S, the resulting output value for six different gain/bias combinations are shown in the remaining columns. Note that for $(\alpha = 1, \beta = 0)$, the input samples are unchanged while for $(\alpha = 1.5, \beta = 0)$ the contrast increases by 50%. Also note that the output samples are truncated to integer values.

5.2.2 Clamping

Table 5.1 also shows that, depending on the gain and bias values, rescaling may produce samples that lie outside of the output image's 8-bit dynamic range. A clamp function is typically used to ensure that the resulting samples remain within an appropriate interval. The clamp function truncates values that are either too small or too large for a specified interval. The clamp function accepts the value to be truncated and an interval, [min, max], to which the input value will be clamped. The clamp function is defined mathematically[1] as

$$clamp(x, min, max) = \begin{cases} min & \text{if } \lfloor x \rfloor \leq min, \\ max & \text{if } \lfloor x \rfloor \geq max, \\ \lfloor x \rfloor & \text{otherwise.} \end{cases}$$

The clamp function is straightforward to implement in Java, as shown in Listing 5.1. Casting the double valued parameter `sample` to an integer effectively computes the floor since this narrowing conversion is equivalent to truncation.

---

[1] The notation $\lfloor X \rfloor$ denotes the floor of X, which is defined as the largest integer that is smaller than or equal to X.

```
1  public class ImagingUtilities {
2    public static int clamp(double sample, int min, int max) {
3      int floorOfSample = (int)sample;
4
5      if (floorOfSample <= min) {
6        return min;
7      } else if (floorOfSample >= max) {
8        return max;
9      } else {
10       return floorOfSample;
11     }
12   }
13 }
```

Listing 5.1. The clamp function.

For an 8-bit image the sample values will typically be clamped to the interval [min=0, max=255] although clients are able to choose the desired range.

The ImagingUtilities class is a collection of static methods commonly used in image processing applications. Since the clamp method is used by many image processing classes we choose to place this static method in the ImagingUtilities class as part of the pixeljelly.utilities package.

The data of Table 5.1 is often depicted graphically as a mapping between input and output values, as shown in the graphs of Figure 5.2. The linear scaling function used for rescaling is a specific type of relation known as a transfer function. A transfer function is any function that relates an input value to an output value in a filtering operation. These graphs illustrate how a linear scaling function relates the input sample, shown on the horizontal axis, to the output sample, which is shown on the vertical axis. The value of the gain determines the slope of the line and hence the contrast of the image. Compare, for example, the upper-left graph with a gain of .25 to the bottom-left, which has a gain of 1.5. If an input image having samples in the range of [0, 150] is processed by these two filters, the output samples would fall into the intervals [0, 37] and [0, 225], respectively. Contrast is controlled by the gain value as seen when comparing the top and bottom rows. Relative brightness is controlled by the bias value as seen by comparing the left and right columns.

Rescaling is perhaps more powerful than might first be apparent. In addition to brightening and darkening an image, rescaling can also invert an image. It is interesting to note that the InvertOp of Listing 4.14 could easily be generalized as a rescaling operation with $\alpha = -1$ and $\beta = 255$.

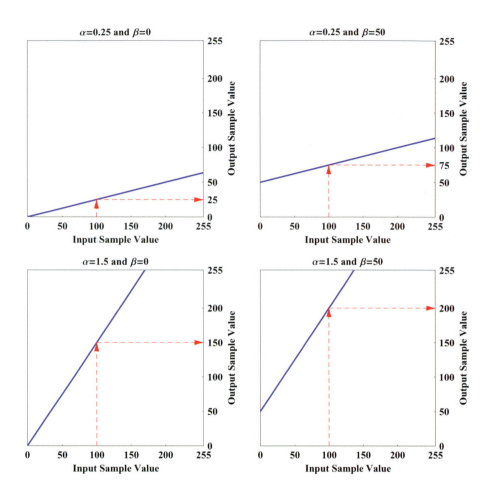

Figure 5.2. Graph of the linear scaling function with various gain and bias settings.

5.2.3   Rescaling Scientific Data for Visualization

Linear scaling is often used to map data that is not essentially visual in nature into an appropriate form for visual analysis. A thermal imaging system, for example, essentially detects *temperatures* by sensing the infrared spectrum. The system may produce a two-dimensional table of temperature readings that indirectly correspond to a visual image but are in the wrong form for rendering on most systems.

Consider a thermal imaging system that takes temperature readings of a black Labrador retriever on a hot summer evening. The imaging system produces,

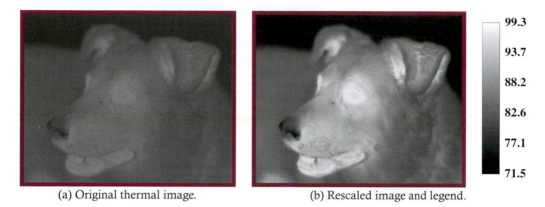

(a) Original thermal image.   (b) Rescaled image and legend.

Figure 5.3. A thermal imaging system captures temperature readings. (Images courtesy of Jason McAdoo.)

in this example, a two-dimensional table of values that correspond to degrees Fahrenheit and range in value from 71.5, the ambient air temperature of the surrounding background, to 99.3, the highest body temperature on the surface of the dog as seen near the eyes and nose of the subject. If these temperatures are directly converted to integer values by rounding they can be treated as a grayscale image, as shown in Figure 5.3(a). The resulting image has very poor contrast and much detail has been lost due to rounding the original data. Instead of transforming the thermal data into an image through rounding, the acquired data should be rescaled in such a way as to make full use of the 8-bit dynamic range so that 71° corresponds to black and 99.3° corresponds to white, as seen in Figure 5.3(b). Rescaling in this way maximizes the contrast of the resulting image and hence preserves as much of the detail as is possible for an 8-bit image.

Contrast Stetching. If the goal of rescaling is to maximize the contrast of an image then a precise selection of the gain and bias parameters is required. Given a two-dimensional data set in the interval $[a, b]$ the optimal gain and bias values for an 8-bit system are given by Equation (5.5). For color depths other than 8 the value 255 should be replaced by the largest sample value allowed in the destination image.

$$
\begin{aligned}
\alpha &= \frac{255}{b-a}, \\
\beta &= -\alpha \cdot \min.
\end{aligned}
\tag{5.5}
$$

When the formulae of Equation (5.5) are applied to the data of the thermal imaging system example, the contrast will be optimized for an 8-bit display when rescaling with the gain and bias values given in Equation (5.6). In this case, the source image is composed of real-valued samples in the range [71.5, 99.3], which are mapped onto the grayscale palette shown to the right of Figure 5.3(b):

$$
\begin{aligned}
\alpha &= \frac{255}{99.3 - 71.5} \\
&= 9.17, \\
\beta &= -\alpha \cdot a \\
&= -9.17 \cdot 71.5 \\
&= -655.85.
\end{aligned}
\tag{5.6}
$$

The contrast stretching presented in Equation (5.5) is problematic when the source image has even a single sample that is either much lower or higher than any other. Rogue samples exert control of the gain in such a way as to seriously diminish the effectiveness of contrast stretching. Alternative techniques seek to reduce the influence of rogue samples by changing the way in which the parameters a and b are chosen. Consider choosing a such that exactly 5% of the source samples are below a and choosing b such that exactly 5% of the source samples are above b. This approach is based on the histogram of an image as defined later in this chapter and serves to significantly reduce the influence of outliers in the source.

5.2.4   Rescaling Color Images

When discussing rescaling we have so far limited ourselves to grayscale images. This process can be naturally extended to color images by rescaling every band of the source using the same gain and bias settings. It is, however, often desirable to apply different gain and bias values to each band of a color image separately.

Consider, for example, a color image that utilizes the HSB color model. Since all color information is contained in the H and S bands, it may be useful to adjust the brightness, encoded in band B, without altering the color of the image in any way. Alternately, consider an RGB image that has, in the process of acquisition, become unbalanced in the color domain. It may be desirable to adjust the relative RGB colors by scaling each band independently of the others. Rescaling the bands of a color image in a nonuniform manner is a straightforward extension to the uniform approach where each band is treated as a single grayscale image.

Examples of both uniform and nonuniform rescaling are shown in Figure 5.4. In part (b) the contrast of the RGB source has been decreased by 70% while in (c) the contrast has increased by 120% (at least prior to clamping) and the brightness increased. In (e) the red band has been kept while the green and blue bands have been effectively turned off by setting the gain of those bands to zero. In (f) the red band has been inverted while the green and blue bands are left unchanged.

(a) Original.

(b) Gain = .7, bias = 0.

(c) Gain = 1.2, bias = 40.

(d) Gain = −1, bias = 255.

(e) Gain = (1,0,0), bias = (0,0,0).

(f) Gain = (−1,1,1), bias = (255,0,0).

Figure 5.4. Illustration of rescaling.

5.2.5   The RescaleOp

Linear scaling is so common in image processing that the Java library contains
the `RescaleOp` class for this purpose. This class is a `BufferedImageOp` implementation of the linear scaling function. The `RescaleOp` can be constructed by
providing two parameters: a float-valued gain and a float-valued bias setting. A
third "rendering hints" parameter must be provided and it is sufficient to specify
a null value. The code of Listing 5.2 shows how to load an image and filter it
with a rescale operator having a gain of 1.6 and a bias of 45. The gain and bias
settings are then uniformly applied to each band of the source.

```
1  BufferedImage  src=ImageIO.read(new  File("image.png"));
2  BufferedImageOp  rescale = new  RescaleOp(1.6f,  45f,  null);
3  BufferedImage  result = rescale.filter(src,  null);
```

Listing 5.2. `RescaleOp` code fragment for uniform scaling.

The `RescaleOp` can also be constructed by providing a separate gain and bias
value for each band of the source image. These gain/bias values are packaged
into float arrays, where the length of each array must be equal to the number of
bands in the source image. The code of Listing 5.3 illustrates how to invert the
red band of a color image while brightening the blue band and increasing the
contrast of the green band.

```
1  BufferedImage  src=ImageIO.read(new  File("image.png"));
2  float[]  gains = {-1f,  0f,  1.1f};
3  float[]  biases = {255f,  35f,  0f};
4  BufferedImageOp  rescale = new  RescaleOp(gains,  biases,  src);
5  BufferedImage  result = rescale.filter(src,  null);
```

Listing 5.3. `RescaleOp` code fragment for nonuniform scaling.

5.3   Lookup Tables

Rescaling can be performed with great efficiency through the use of lookup tables, assuming that the color depth of the source is sufficiently small. Consider,
for example, rescaling an 8-bit image. Without using lookup tables we compute
the value $\mathrm{clamp}(\alpha \cdot S + \beta, 0, 255)$ for every sample S in the input. For an $W \times H$
image there are WH samples in the input and each of the corresponding output
samples requires one multiplication, one addition, and one function call.

```
1  algorithm  rescaleOp (Image  input ,  Image  output ,  float  gain ,  float  bias )
2      int []  lookupTable  =  new  int [256];
3      for  S  =  0  to  255
4          lookupTable [S]  =  clamp (gain  *  S  +  bias ,  0,  255);
5
6      for  every  sample  S  in  the  input
7          set  the  corresponding  destination  sample  to  lookupTable [S]
```

Listing 5.4. Informal description of lookup tables and rescaling.

But note that since every sample in the source must be in the range $[0, 255]$ we need only compute the 256 possible outputs exactly once and then refer to those precomputed outputs as we scan the image. In other words, computational effort can be reduced by computing the output for each *possible* input sample rather than each *actual* input sample. In an 8-bit image, for example, each sample will be in the range $[0, 255]$. Since there are only 256 possible sample values for which the quantity $\mathrm{clamp}(\alpha \cdot S + \beta, 0, 255)$ must be computed, time can be saved by precomputing each of these values, rather than repeatedly computing the scaled output for each of the WH actual input samples.

The precomputed mapping between input and output samples can be represented as an array known as a lookup table. In a lookup table, the indices represent the input sample values and each array cell stores the corresponding output value. Listing 5.4 presents pseudocode that rescales an image by utilizing a lookup table assuming the output image uses 8-bit samples.

In order to illustrate the computational savings that lookup tables provide, consider that for an 8-bit grayscale image the number of floating point arithmetic operations (a single multiplication and addition excluding consideration of the clamping function) is reduced from $2 \cdot W \cdot H$ to $2 \cdot 256$. The total number of arithmetic operations saved by using a lookup table to rescale an 8-bit grayscale image that is 640 pixels wide by 480 pixels high is then $(2 \cdot 640 \cdot 480 - 2 \cdot 256)$ or 613,888 arithmetic operations!

While lookup tables will generally improve the computational efficiency of rescaling,[2] they are also more flexible. The rescaling operation represents a linear relationship between input and output sample values but a lookup table is able to represent both linear and nonlinear relationships of arbitrary complexity.

The LookupOp class contained in Java's image processing library efficiently supports linear and nonlinear point-processing operations. The LookupOp is a BufferedImageOp that must be constructed by providing a LookupTable object

---

[2]Some image processing filters may not benefit from lookup tables if the operations are computationally simple. Integer-based arithmetic operations may outperform lookup tables.

```
1  // define the lookup table
2  byte[] table = new byte[256];
3  for(int i=0; i<table.length; i++) {
4      table[i] = (byte)clamp(gain*i+bias, 0, 255);
5  }
6
7  BufferedImageOp lookup = new LookupOp(new ByteLookupTable(0, table), null);
8  BufferedImage src=ImageIO.read(new File("image.png"));
9  BufferedImage result = lookup.filter(src, null);
```

Listing 5.5. Code fragment using LookupOp and LookupTable.

as a parameter. The ByteLookupTable is a LookupTable that is constructed by passing in an array of bytes and an offset value that is subtracted from all entries in the table. The code of Listing 5.5 shows how to construct a LookupOp that is functionally equivalent to the RescaleOp discussed earlier.

5.4   Gamma Correction

Gamma correction is an image enhancement operation that seeks to maintain perceptually uniform sample values throughout an entire imaging pipeline. In general terms, an image is processed by (1) acquiring an image with a camera, (2) compressing the image into digital form, (3) transmitting the data, (4) decoding the compressed data, and (5) displaying it on an output device. Images in such a process should ideally be displayed on the output device exactly as they appear to the camera. Since each phase of the process described above may introduce distortions of the image, however, it can be difficult to achieve precise uniformity.

Gamma correction seeks to eliminate distortions introduced by the first and the final phases of the image processing pipeline. Both cameras and displays (e.g., computer monitors, projectors) possess intrinsic electronic properties that introduce nonlinear distortions in the images that they produce. For a display, this nonlinearity is best described as a power-law relationship between the level of light that is *actually* produced by the display and the amount of signal that controls the display. Stated another way, we might say that there is a power-law relationship between the level of light that is actually produced and the amount of light that is *intended* for display. The power-law relationship between the displayed and intended light levels is given by Equation (5.7):

$$\hat{S}_{\text{displayed}} = (\hat{S}_{\text{intended}})^{\gamma}. \tag{5.7}$$

This formulation assumes all samples are normalized into the interval $[0, 1]$ as indicated by the hat (a.k.a. "circumflex") notation. Additionally, note that gamma, the Greek symbol γ, is an exponent that determines the amount of distortion produced by a particular output device, where a gamma value of 1 indicates that no distortion is introduced and all others indicate some degree of distortion. Computer monitors and televisions typically have gamma values ranging from 1.2 to 2.5. Output devices are often characterized by their gamma values, which are usually supplied by the device manufacturer.

When applied to image processing applications, this power-law states that the amount of light actually displayed for a sample is a slight distortion of the amount of light intended for display. Since most output devices distort an image according to this power-law, it is necessary to remove this distortion by a process called gamma correction. Since the output device intrinsically introduces some distortion, the distortion must be corrected via software. Given a displays gamma value, intended sample values are raised to the power of $1/\gamma$ prior to display which cancels any distortion caused by the output device. Gamma correction is shown in Equation (5.8):

$$\hat{S}_{\text{corrected}} = (\hat{S}_{\text{intended}})^{1/\gamma}. \tag{5.8}$$

Once the intended sample is gamma-corrected it is then displayed on the output device. Substitution of the gamma corrected sample from Equation (5.8) into the output device of Equation (5.7) shows how gamma correction eliminates the distortion caused by the display hardware. This substitution and the ensuing simplification is shown in Equation (5.9):

$$\begin{aligned} \hat{S}_{\text{displayed}} &= (\hat{S}_{\text{corrected}})^{\gamma}, \\ \hat{S}_{\text{displayed}} &= \left((\hat{S}_{\text{intended}})^{1/\gamma}\right)^{\gamma}, \\ \hat{S}_{\text{displayed}} &= \hat{S}_{\text{intended}}. \end{aligned} \tag{5.9}$$

Gamma correction is often explicitly encoded in image file formats. The PNG specification, for example, defines a gAMA[3] chunk (a portion of the image data) that "specifies the relationship between the image samples and the desired display output intensity." This is important because an image that is created or edited using an image processing application on one computer system may not look the same when viewed on another system if the displays have differing gamma values. PNG encodes the gamma value of the display on which the image was initially created and this value can then be used by PNG-aware software to adjust the decoded image for display on output devices having different display properties.

---

[3] Although the name appears improperly capitalized, gAMA is the correct name for the gamma information in a PNG file.

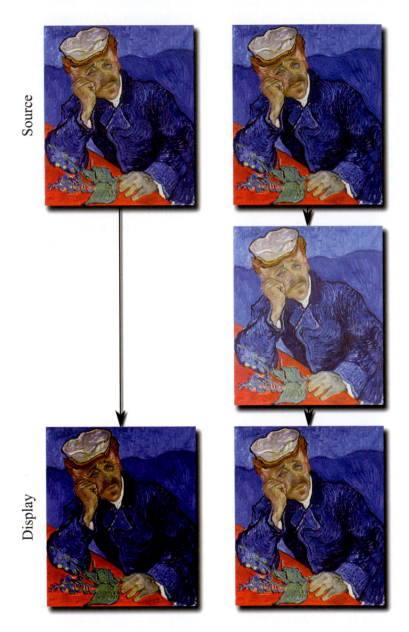

Figure 5.5. Gamma correction, as shown in the middle of the column on the right.

Figure 5.5 illustrates how gamma correction negates the distortions (typically over- darkening) introduced by a display device. Two separate pipelines are illustrated, where the leftmost pipeline does not gamma correct but the rightmost pipeline is properly gamma corrected. In each case, the uppermost image represents a source as it should appear on the display. The center portion of the pipeline allows for gamma correction that is omitted in the leftmost system but corrected in the rightmost pipeline, where the display is known to have a gamma value of 2.5. The bottom row reflects how the source image is overly darkened by the leftmost display since it is not corrected, while the rightmost image is accurately rendered.

Gamma correction is a nonlinear transformation that can be implemented by subclassing the LookupOp, as shown in Listing 5.6. The constructor accepts a gamma value and a Boolean flag indicating whether to apply *gamma correction* or to simulate *gamma distortion*. The constructor calls a static method that builds the lookup table that is then passed to the superclasse's constructor. This implementation works exclusively on 8-bit images since the lookup table contains only 256 entries. It should also be noted that indexed color models are not supported.

```
1  public class GammaOp extends LookupOp {
2    private static ByteLookupTable getTable(double gamma, boolean invert) {
3      if(invert) gamma = 1/gamma;
4      byte[] table = new byte[256];
5      for(int i=0; i<table.length; i++) {
6        table[i] = (byte)(255 * Math.pow(i/255.0, gamma));
7      }
8
9      return new ByteLookupTable(0, table);
10   }
11
12   public GammaOp(double gamma, boolean invert) {
13     super(getTable(gamma, invert), null);
14   }
15 }
```

Listing 5.6. The GammaOp class.

5.5 Pseudo Coloring

A pseudo-colored image is one that renders its subject using a coloring scheme that differs from the way the subject is naturally perceived. While the term *pseudo* signifies false or counterfeit, a pseudo-colored image should not be understood as an image that is defective or misleading but as an image that is colored in an unnatural way. Pseudo coloring is a point-processing technique that generates a pseudo-colored image from a typically grayscale source.

Many scientific instruments take spatial measurements of nonvisual data that are then rendered for visual interpretation and analysis. Pseudo-colored images are often used to visualize information gathered by instruments measuring data outside of the visible spectrum. Satellite measurements of rainfall amount or earth surface temperature readings are, for example, spatially distributed scalar measures that can be rendered as an image. The data is typically scalar (i.e., grayscale) and, since it is not intrinsically visual, may possess poor qualities when directly rendered as an image.

While rescaling is able to maximize grayscale contrast, it is often desirable to render the image using a color scheme that further enhances the visual interpretation of the data. A thermal imaging system may, for example, record surface temperatures of a subject and produce data over a range of degrees Fahrenheit. This data can be rescaled, following the process described in the previous section to maximize the grayscale contrast, as shown in image (a) of Figure 5.6. The visual contrast can, however, be further augmented by artificially applying colors to specific temperature values. Since blue is generally associated with cooler temperatures and yellow with hotter temperatures, a color scheme that ranges from blue to yellow can be artificially applied to the temperature data, as illustrated in image (b) of Figure 5.6. The correspondence between measured data (temperature values) and colors is given in the color bar to the right of the images.

The pseudo-coloring operation can best be implemented by building a customized `BufferedImageOp`, where the filter method accepts a grayscale image and produces a color image having an `IndexColorModel`. Recall that the samples of an indexed image represent indices into a color palette and the `IndexColorModel` is a good way of representing the palette.

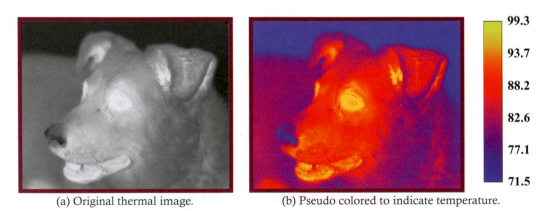

(a) Original thermal image.   (b) Pseudo colored to indicate temperature.

Figure 5.6. False coloring example.

```
1  byte[] reds, blues, greens;
2  // Initialization of these arrays is not shown but is assumed.
3  // All three arrays are equal length.
4
5  IndexColorModel model = new IndexColorModel(8, reds.length, reds, greens, blues);
6  BufferedImage image =
7      new BufferedImage(640, 480, BufferedImage.TYPE_BYTE_INDEXED, model);
```

Listing 5.7. Code fragment of the IndexColorModel.

```
1  public class FalseColorOp extends NullOp {
2    protected byte[] reds, greens, blues;
3
4    public FalseColorOp(Color[] pallette) {
5      reds = new byte[pallette.length];
6      greens = new byte[pallette.length];
7      blues = new byte[pallette.length];
8
9      for(int i=0; i<reds.length; i++){
10        reds[i] = (byte)pallette[i].getRed();
11        greens[i] = (byte)pallette[i].getGreen();
12        blues[i] = (byte)pallette[i].getBlue();
13      }
14    }
15
16    private IndexColorModel getColorModel() {
17      return new IndexColorModel(8, reds.length, reds, greens, blues);
18    }
19
20    public BufferedImage createCompatibleDestImage(BufferedImage src,
21                                                   ColorModel destCM) {
22      return new BufferedImage(src.getWidth(),
23                              src.getHeight(),
24                              BufferedImage.TYPE_BYTE_INDEXED,
25                              getColorModel());
26    }
27
28    public BufferedImage filter(BufferedImage src, BufferedImage dest) {
29      if(dest == null) dest = createCompatibleDestImage(src, null);
30
31      return super.filter(src, dest);
32    }
33  }
```

Listing 5.8. Implementation of the FalseColorOp.

An IndexColorModel object can be constructed by specifying three byte-valued arrays containing the red, green, and blue bands of the color palette, respectively. In addition, the number of bits per band (typically 8) and the number of colors in the palette (typically 256) must also be supplied. Once the IndexColorModel has been constructed, it can be used to generate a Buffered Image. The code fragment of Listing 5.7 shows how to construct a true-color 640 by 480 pixel image using this technique.

An implementation of the pseudo-coloring operation is shown in Listing 5.8. Note that the FalseColorOp constructor accepts an array of Color objects corresponding to the colors in the palette and splits the array into its respective red, green, and blue bands. The createCompatibleDestImage method is overridden to construct a BufferedImage that contains the appropriate indexed color model and uses byte-valued samples that are treated as indices into the color palette. Data from the grayscale source is then directly copied into the destination raster by the default NullOp filter method.

Pixel Jelly supports pseudo coloring. The colorizing operation allows users to construct a pallette, which is often referred to as gradient, by dragging and dropping colors onto the gradient editor. The gradient is then applied to the source image in order to colorize the source. If the source image is a color image the application first converts the source to grayscale after which the gradient is applied. Figure 5.7 shows a screenshot of this operation, where the color source image is given in part (a) and the false-colored result is shown in part (b).

(a) Original.   (b) Pseudo coloring.

Figure 5.7. False coloring example.

5.6 Histogram Equalization

Histogram equalization is a powerful point processing enhancement technique that seeks to optimize the contrast of an image. As the name of this technique suggests, histogram equalization seeks to improve image contrast by flattening, or equalizing, the histogram of an image.

A histogram is a table that simply counts the number of times a value appears in some data set. In image processing, a histogram is a histogram of sample values. For an 8-bit image there will be 256 possible samples in the image and the histogram will simply count the number of times that each sample actually occurs in the image.

Consider, for example, an 8-bit $W \times H$ grayscale image. There are 256 distinct sample values that *could* occur in the image. The histogram of the image is a table of 256 values where the ith entry in the histogram table contains the number of times a sample of value i occurs in the image. If the image were entirely black, for example, the 0th entry in the table would contain a value of $W \times H$ (since all the image pixels are black) and all other table entries would be zero. In general, for an N-bit $W \times H$ grayscale image where the ith sample is known to occur n_i times, the histogram h is formally defined by Equation (5.10):

$$h(i) = n_i \quad i \in 0, 1, \ldots, 2^N \tag{5.10}$$

Histograms are typically normalized such that the histogram values sum to 1. In Equation (5.10) the histogram is not normalized since the sum of the histogram values is WH. The normalized histogram is given in Equation (5.11), where $\hat{h}(i)$ represents the probability that a randomly selected sample of the image that will have a value of i:

$$\hat{h}(i) = h(i)/(WH) = n_i/(WH), \quad i \in 0, 1, \ldots, 2^N. \tag{5.11}$$

A histogram is typically plotted as a bar chart where the horizontal axis corresponds to the dynamic range of the image and the height of each bar corresponds to the sample count or the probability. Generally, the overall shape of a histogram doesn't convey much useful information but there are several key insights that can be gained. The spread of the histogram relates directly to image contrast where narrow histogram distributions are representative of low contrast images while wide distributions are representative of higher contrast images. Generally, the histogram of an underexposed image will have a relatively narrow distribution with a peak that is significantly shifted to the left while the histogram of an overexposed image will have a relatively narrow distribution with a peak that is significantly shifted to the right.

Figure 5.8 shows a relatively dark grayscale image and its corresponding histogram. The histogram is shifted to the left and has a relatively narrow distribution since most of the samples fall within the narrow range of approximately 25

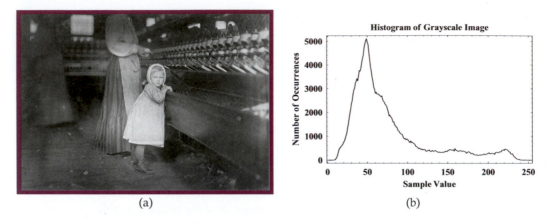

(a)   (b)

Figure 5.8. An example histogram: (a) an 8-bit grayscale image and (b) its histogram.

to 80, while relatively few of the images samples are brighter than 128. The histogram for this example is indicative of an underexposed image—one that *may* be improved through histogram equalization.

Histogram equalization is a way of improving the local contrast of an image without altering the global contrast to a significant degree. This method is especially useful in images having large regions of similar tone such as an image with a very light background and dark foreground. Histogram equalization can expose hidden details in an image by stretching out the contrast of local regions and hence making the differences in the region more pronounced and visible.

Equalization is a nonlinear point processing technique that attempts to map the input samples to output samples in such a way that there are equal amounts of each sample in the output. Since equalization is a point processing technique it is typically implemented through the use of a lookup table. For a source image, equalization computes the histogram of the source and then constructs a discrete cumulative distribution function (CDF) which is used as the lookup table. Given an N-bit image having histogram h, the normalized CDF $\hat{C}$ is defined in Equation (5.12):

$$\hat{C}_j = \sum_{i=0}^{j} \hat{h}_i, \quad j \in \{0, 1, \ldots, 255\}. \tag{5.12}$$

The cumulative distribution function essentially answers the question, "What percentage of the samples in an image are equal to or less than value J in the image?" Since the resulting CDF lies in the interval $[0 \ldots 1]$ it can't be directly used as a lookup table since the output samples must lie in the 8-bit range of $[0 \ldots 255]$. The CDF is therefore scaled to the dynamic range of the output image. For an 8-bit image, each entry of the CDF is multiplied by 255 and rounded to obtain

i	$\hat{h}_i$	$\hat{C}_i$	$7\hat{C}_i$
0	1/16	1/16	0
1	3/16	4/16	2
2	4/16	8/16	4
3	4/16	12/16	5
4	3/16	15/16	7
5	1/16	16/16	7
6	0/16	16/16	7
7	0/16	16/16	7

(a) image:

0	1	3	4
1	2	2	3
1	3	4	4
3	2	5	2

(c) equalized image:

0	2	5	7
2	4	4	5
2	5	7	7
5	4	7	4

(d) histogram:

i	$\hat{h}_i$
0	1/16
1	0/16
2	3/16
3	0/16
4	4/16
5	4/16
6	0/16
7	4/16

(a)   (b)   (c)   (d)

Figure 5.9. Numerical example of histogram equalization: (a) a 3-bit image, (b) normalized histogram and CDF, (c) the equalized image, and (d) histogram of the result.

the lookup table used in histogram equalization. In summary then, histogram equalization works by (1) computing the histogram of the source image, (2) generating the CDF of the source image, (3) scaling the CDF to the dynamic range of the output and (4) using the scaled CDF as a lookup table.

Figure 5.9 gives a numeric example of an underexposed image that is equalized to improve the local contrast. In this figure the 3-bit grayscale image shown in (a) is used to generate the normalized histogram and CDF shown in (b). Since the source is a 3 bit image, the CDF is scaled by a factor of seven, the dynamic range of the 3-bit output, to produce the lookup table used for equalization. In this example, then, every 0 in the input remains 0 in the output while every 1 in the input becomes a 2 in the output and so on. The resulting equalized image is shown in (c) and the histogram of the output is shown in (d). While the resulting histogram is not *exactly* equal at every index, the resulting histogram spreads the samples across the full dynamic range of the result and does increase local contrast in the image.

Figure 5.10 shows an underexposed grayscale image in (a) that has the histogram shown in (b). The underexposed image is histogram-equalized in (c) such that the samples are nearly uniformly distributed across the 8-bit range. The histogram of the equalized image is shown in (d). The net effect of equalization in this case is to shift the darker regions upwards in tone thus brightening much of the image.

Images with poor overall contrast can also be corrected using histogram equalization, as shown in Figure 5.11. In this case, the image 8-bit source image of (a) is neither overly dark nor overly bright but has a washed-out look since most of the samples are in a narrow range of middle gray tone. There are no strong bright or dark regions in the image as indicated by the histogram in (b). The equalized image of (c) presents more stark edges (i.e., differences between sample values are increased) and the histogram of the resulting image has been generally flattened, as shown in (d).

(a) Underexposed image.

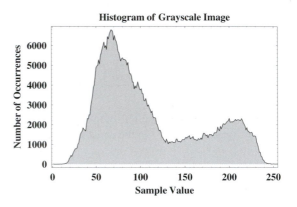

(b) Histogram of the underexposed image.

(c) Histogram-equalized image.

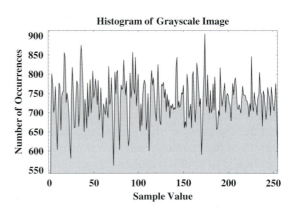

(d) Histogram of the equalized image.

Figure 5.10. Histogram equalization of an underexposed image.

(a) Low-contrast image.

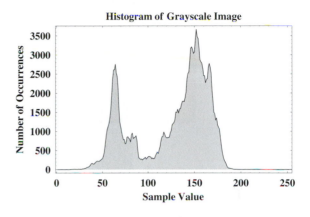

(b) Histogram of the low-contrast image.

(c) Histogram equalized image.

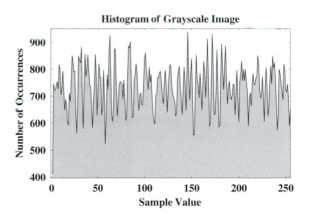

(d) Histogram of the equalized image.

Figure 5.11. Histogram equalization of a low-contrast image.

(a) Source.   (b) Equalized RGB.   (c) Equalized intensity.

Figure 5.12. Equalizing a color image.

Histogram equalization can also be done on color images by performing the grayscale technique on each separate band of the image. Care should be taken when doing this, however, since the colors will likely be dramatically altered as a result. If the tonal distributions are different among the red, green, and blue channels of an image, for example, the lookup tables for each channel will be vastly different and equalization will alter the color patterns present in the source. Histogram equalization of a color image is best performed on the intensity channel only, which implies that the equalization should be done on the brightness band of an image using the HSB or YIQ color spaces, for example.

Figure 5.12 demonstrates how equalization may alter the color balance of a source image. In this figure, the source image has a largely reddish tint. The histogram of the red channel is shifted to the high end while the histogram of the green channel is shifted to the low. When equalization is done on each of the red, green, and blue channels independently, the color balance is dramatically altered as seen in (b). When equalization is done on the intensity channel only, the chromaticity is retained but the brightness levels are altered, as shown in (c).

Since image histograms are so useful in image processing it is advisable to develop a `Histogram` class for the purpose of creating and processing histogram data. Listing 5.9 shows such a class, where the constructor accepts a `BufferedImage` source and the band from which to extract the histogram. The constructor then determines the maximum possible sample value and allocates a table of the proper size. The table is filled in by scanning the source and counting each sample that occurs in the image.

A histogram object can convert itself to an array via the `getCounts` methods, it can produce a cumulative distribution function as an array of doubles via the `getCDF` method, and it can produce a normalized histogram via the `getNormalizedHistogram` method. The histogram class can be used to implement a `HistogramEqualizationOp`, which is left as a programming exercise.

```
1  public class Histogram {
2    private int[] counts;
3    private int totalSamples, maxPossibleSampleValue;
4
5    public Histogram(BufferedImage src, int band) {
6      maxPossibleSampleValue = getMaxSampleValue(src, band);
7      counts = new int[maxPossibleSampleValue + 1];
8      totalSamples = src.getWidth() * src.getHeight();
9
10     for (Location pt : new RasterScanner(src, false)) {
11       int sample = src.getRaster().getSample(pt.col, pt.row, band);
12       counts[sample]++;
13     }
14   }
15
16   public int getNumberOfBins() {
17     return counts.length;
18   }
19
20   public int getValue(int index){
21     return counts[index];
22   }
23
24   public int[] getCounts() {
25     int[] result = new int[counts.length];
26     System.arraycopy(counts, 0, result, 0, counts.length);
27     return result;
28   }
29
30   public double[] getCDF() {
31     double[] cdf = getNormalizedHistogram();
32     for(int i=1; i<cdf.length; i++){
33       cdf[i] = cdf[i-1] + cdf[i];
34     }
35     return cdf;
36   }
37
38   public double[] getNormalizedHistogram() {
39     double[] result = new double[counts.length];
40     for(int i=0; i<counts.length; i++){
41       result[i] = counts[i] / (double)totalSamples;
42     }
43     return result;
44   }
45
46   private int getMaxSampleValue(BufferedImage src, int band) {
47     return (int)Math.pow(2, src.getSampleModel().getSampleSize(band)) - 1;
48   }
49 }
```

Listing 5.9. Histogram class.

Histograms have thus far been presented as one-dimensional constructs that process each channel of an image independently. Color histograms, by contrast, are three dimensional constructs that divide the images color space into volumetric bins. Each bin encloses a rectangular volume of colors in the space and the corresponding bin entry is a count of the number of pixels that are enclosed by the bin. A color histogram correctly links the individual samples of a pixel together in a dependent fashion.

The size of the bins determines the resolution of the histogram, where smaller bins correspond to a greater resolution and larger bin sizes correspond to a lower resolution. While higher resolution histograms provide more accuracy in terms of characterizing the color distribution of an image, higher resolutions also lead to larger memory consumption and a corresponding lack of computational efficiency. Consider, for example, an 8-bit RGB color image such that each axis is divided into 256 bins. The color histogram has an impractical total of $256 \times 256 \times 256 = 16,777,216$ bins. The resolution of the histogram can be reduced by dividing each axis into only 5 bins such that the color histogram then has a computationally effective total of $5 \times 5 \times 5 = 125$ bins. Each axis of the color space may be given a resolution independently of the others such that the Y axis of the YCbCr color space may be given a higher resolution than either the Cb or Cr color spaces. Consider, for example, an 8-bit YCbCr image such that the Y axis is divided into 10 bins while each of the Cb and Cr axes are divided into 5 bins. The corresponding color histogram has a total of $10 \times 5 \times 5 = 250$ bins and has superior resolution in the intensity channel than the two chroma channels.

Figure 5.13 shows the RGB color histogram for two source images. The $12 \times 12 \times 12$ color histogram of part (a) is shown in (b) where the diameter of the spheres is directly proportional to the count of each volumetric bin and the color of each sphere corresponds to the color at the center of each bin. The $12 \times 12 \times 12$ color histogram of the source image shown in part (c) is given in (d). The large density of dark and red pixels of the source image in (a) is reflected in the distribution of data in the color histogram of (b) while the large density of light blue pixels is reflected in the distribution of the color histogram of part (d).

Since a color histogram provides a relatively coarse but concise characterization of an image, a color histogram can be used as a computationally effective means for comparing the similarity of two images. Color histograms are often used in content-based image retrieval (CBIR) systems to support efficient searches of large image databases. A CBIR system maintains a database of images that can be queried for similarity to a target image. A CBIR system allows users to find images in the database that are similar to the target image rather than searching the image database using metadata such as keywords. While two visually different images may have similar color histograms, the color histograms can be used as a coarse measure of similarity and serve as a preprocessing step in a CBIR query.

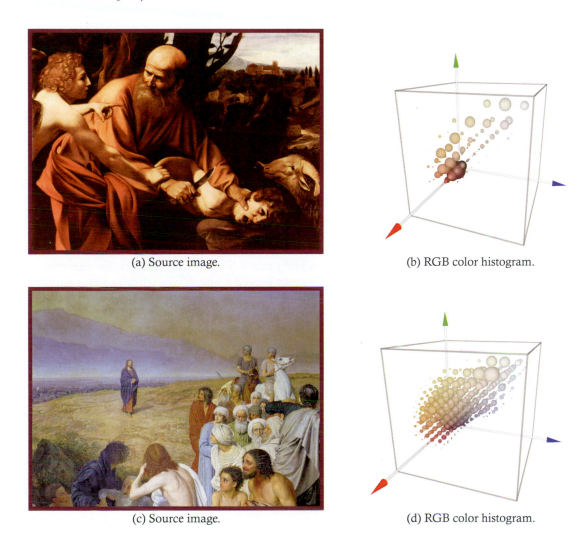

(a) Source image.

(b) RGB color histogram.

(c) Source image.

(d) RGB color histogram.

Figure 5.13. The color distribution of two source images as given by their $12 \times 12 \times 12$ RGB color histograms.

5.7 Arithmetic Image Operations

Two source images can be added, multiplied, one subtracted from the other or one divided by the other to produce a single destination. These operations are known as image arithmetic or image blending. Each of these methods is a point processing operation where corresponding samples from the two source images are combined into the destination. Image addition, for example, works by summing corresponding samples from both source images to produce the resulting sample. Image addition is the basis for many blending modes used in image processing software as well as in computer graphics rendering pipelines. Both graphic designers and software engineers are well advised to understand the underlying principles of image addition presented in this section.

It is important to note that since the operations presented in this section combine *multiple* source images into a single destination, these operations are not single source filters. Since Java's `BufferedImageOp` interface is designed for single source operations, the implementation of the arithmetic operations, although straightforward, is somewhat awkward when integrated into the single source framework of the `BufferedImageOp`.

Before looking at implementation, however, we will discuss two complications that arise when performing image arithmetic. The first relates to the color depth of the destination and the second relates to the dimension of the destination. Consider that when adding two 8-bit grayscale images, for example, the sum of the two input samples may exceed 8 bits. If two white samples are added together the result will be $255 + 255 = 510$, which is a 9-bit quantity. This would imply that color depth of the destination will generally be larger than the color depth of the source. Rather than increasing the color depth of the destination, however, most implementations will simply rescale the destination into an 8-bit range by taking the average of the two inputs. Another alternative is to simply clamp the result to an 8-bit range; although this approach will often truncate so much information that it is not often used.

The second complication arises if the source images are not of the same width and height. Given this scenario, image addition is carefully defined to either forbid the operation or to modify the sources such that they are of identical dimension. Given source images I_a and I_b of dimension $W_a \times H_a$ and $W_b \times H_b$, respectively, the sum $I_a + I_b$ produces destination image I_{dst}, which is as wide as the narrowest of the two sources and as high as the shortest of the two sources. Addition with scaling is formally given as

$$I_{dst}(x, y) = \frac{1}{2} \cdot (I_a(x, y) + I_b(x, y)), \tag{5.13}$$

where the dimensions of I_c are $\min(W_a, W_b) \times \min(H_a, H_b)$ and variables x and y range over all column and row indices. To simplify our presentation we define s_1 as the sample given by $I_a(x, y)$, s_2 as $I_b(x, y)$ and finally s_{dst} as the destination sample, $I_c(x, y)$. Making these substitutions into Equation (5.13) yields the more convenient

$$s_{dst} = \frac{1}{2} \cdot (s_1 + s_2), \tag{5.14}$$

where the destination sample is computed over all samples of the two sources.

(a) I_a.   (b) I_b.   (c) $I_a + I_b$.

Figure 5.14. Image addition.

Addition with clamping is given as

$$s_{\text{dst}} = clamp(s_1 + s_2,\ 0,\ 255).$$

Figure 5.14 shows an example of image addition. In this example an 8-bit grayscale image is duplicated and horizontally flipped. The original and flipped sources are then added to form the result. The destination image has been rescaled into an 8-bit color depth rather than clipped.

Consider multiplying a full color source with an image having only white and black. The binary image is known as a mask where the *white* pixels simply pass the source image through to the output unaltered while the black pixels erase the corresponding source images pixels. Image multiplication is a good technique for either adding color to a line drawing or for emphasizing edges in a full color image. Figure 5.15 illustrates image multiplication. In part (a) of this figure, a source image is multiplied by the mask of (b) to obtain the output of (c). Many image processing applications use image multiplication to create artistic filters for digital artists.

Image subtraction operates on two sources by subtracting corresponding sample values. The dimensions of the resulting image are determined in precisely the same fashion as image addition, where the resulting dimensions represent the smallest region common to both source images. The color depth of the resulting image is also problematic since the resulting image may contain both positive and negative values. The color depth problem can be resolved using the same techniques employed in image addition or, as is more commonly done, through taking the absolute value of the output. Image subtraction is typically defined as

$$s_{\text{dst}} = |s_1 - s_2|.$$

(a) I_a.   (b) I_b.   (c) $I_a \cdot I_b$.

Figure 5.15. Image multiplication.

Image subtraction is useful for visualizing the difference between two images. One way of telling whether two images are identical is to subtract one from the other and if the maximum sample value of the result is zero then the images are identical. Also, the destination image gives a visualization of the differences such that areas of difference can be clearly seen. Figure 5.16 shows a source

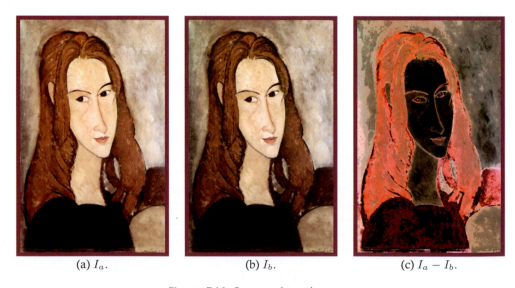

(a) I_a.   (b) I_b.   (c) $I_a - I_b$.

Figure 5.16. Image subtraction.

image (a) that has been histogram equalized on the intensity channel to obtain the result shown in (b). In order to visualize how equalization affected various parts of the source we can subtract the equalized image from the source to obtain the image of (c), which has been rescaled to maximize the image contrast. In this example, the brighter regions correspond to areas of greater difference while the darker regions correspond to areas where equalization had little effect.

5.7.1   BinaryImageOp Class

The `BufferedImageOp` class is not well designed to represent the arithmetic operations since there are two source images. Nonetheless we can implement image arithmetic as a single source operation by currying. Currying refers to the transformation of a function that has multiple arguments into a function that has only a single argument. For our implementation, we develop an abstract base class called `BinaryImageOp` that is partially shown in Listing 5.10. In this class, the filter method iterates over corresponding samples of the two sources and blends them using the abstract `combine` function. This function accepts the two source samples and combines them using either the arithmetic or logic operators. It is also very straightforward to combine the two inputs in nonlinear ways such as computing the maximum or minimum value. The `combine` method as given in this class is left abstract, however, since the specific operation that will be applied is known only by the implementing subclass.

In order to construct a binary image operation the client must provide the *first* or *leftmost* image of the two images to be combined. This image is then saved as an instance variable and is used by the filter method whenever the operation is applied to the second source. The filter method should then be understood as meaning *combine the leftmost image with the rightmost image that is now supplied as the source*. By caching the leftmost source image we have essentially created a curried version of the filter method that allows us to support multiple source operations via the otherwise single-source `BufferedImageOp` interface.

The implementation of the filter method involves some overhead to adequately deal with source images that may be of differing dimensions or color depths. The initial lines of the filter method determine the dimensions of the destination. The `Rectangle` class is part of the standard Java distribution and provides a convenient way for computing the smallest of two rectangles, the boundaries of the two source images, through the `intersection` method. In addition, the `createCompatibleDestination` method is rewritten to accept the desired bounding `Rectangle` of the destination and the desired color model of the destination and create a suitable destination image. The number of bands of the destination will be the minimum of the number of bands of the two sources.

```
 1  public abstract class BinaryImageOp extends NullOp {
 2    protected BufferedImage left;
 3
 4    public BinaryImageOp(BufferedImage left) {
 5      this(left);
 6    }
 7
 8    public abstract int combine(int s1, int s2);
 9
10    public BufferedImage createCompatibleDestImage(Rectangle bounds,
11                                                    ColorModel destCM) {
12      return new BufferedImage(
13        destCM,
14        destCM.createCompatibleWritableRaster(bounds.width, bounds.height),
15        destCM.isAlphaPremultiplied(),
16        null);
17    }
18
19    public BufferedImage filter(BufferedImage right, BufferedImage dest) {
20      Rectangle leftBounds = left.getRaster().getBounds();
21      Rectangle rightBounds = right.getRaster().getBounds();
22      Rectangle intersection = leftBounds.intersection(rightBounds);
23
24      if (dest == null) {
25        if (left.getRaster().getNumBands() < right.getRaster().getNumBands()) {
26          dest = createCompatibleDestImage(intersection, left.getColorModel());
27        } else {
28          dest = createCompatibleDestImage(intersection, right.getColorModel());
29        }
30      }
31
32      for(Location pt : new RasterScanner(dest, true)) {
33        int s1 = left.getRaster().getSample(pt.col, pt.row, pt.band);
34        int s2 = right.getRaster().getSample(pt.col, pt.row, pt.band);
35        dest.getRaster().setSample(pt.col, pt.row, pt.band, combine(s1, s2));
36      }
37      return dest;
38    }
39  }
```

Listing 5.10. `BinaryImageOp`.

The `BufferedImageOp` is abstract since we define a `combine` method that takes two sample values and combines them in a way that known only by the implementing subclass. Implementing the various arithmetic ops becomes a very simple matter of writing a concrete subclass and providing an implementation of the `combine` method. Listing 5.11 shows a complete implementation of the `BinaryImageOp` that supports image addition. The `combine` method takes the two source samples given by s_1 and s_2 and adds them together, rescales them by dividing by 2, and then clamps them to an 8-bit value. This class can be followed as a pattern for implementing the various other binary image operations described. In order to create an image subtraction operation, a `SubtractBinaryOp` class should be developed that provides a constructor and a `combine` method similar to the one shown here for addition.

```java
public class AddBinaryOp extends BinaryImageOp {
    public AddBinaryOp(BufferedImage left){
        super(left);
    }

    public int combine(int s1, int s2){
        return ImagingUtilities.clamp((s1 + s2)/2.0, 0, 255);
    }
}
```

Listing 5.11. `AddBinaryOp`.

5.8   Logical Image Operations

Binary images can also be combined using logical operators known as conjunction (AND), disjunction (OR), negation (NOT), or exclusive or (XOR). Under these logical operations, each sample is implicitly understood as either True (white) or False (black). Logical negation is a unary operator having a single binary source image that is inverted; meaning that all white pixels become black and all black pixels become white. The remaining logical operators operate on precisely two binary sources and hence are not single source filters.

Given two binary input samples as input there are only four cases that these operators need to consider: namely (black, black), (black, white), (white, black), and (white, white). Table 5.2 gives each of these cases and defines the output

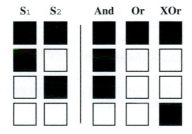

Table 5.2. Truth table for logical operators on binary images.

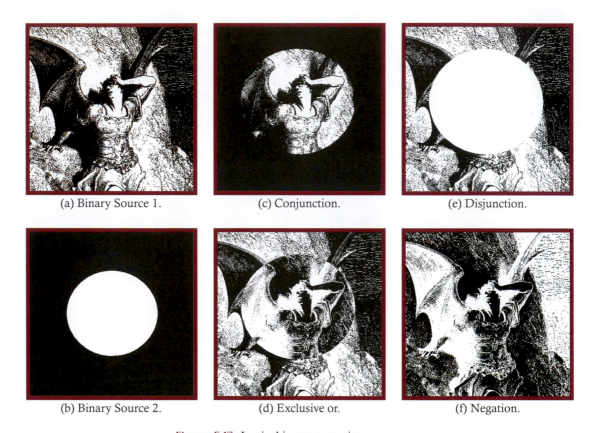

(a) Binary Source 1.   (c) Conjunction.   (e) Disjunction.

(b) Binary Source 2.   (d) Exclusive or.   (f) Negation.

Figure 5.17. Logical image operations.

under each operator. Columns S_1 and S_2 represent the two source samples while the columns to the right show the output of each of the three logical operators. The logical conjunction of two black samples, for instance, results in a black output while the logical conjunction of two white samples results in a white output.

Figure 5.17 illustrates the logical operations as applied on two binary sources. The white circle on black background serves as a mask that either copies everything inside of its boundary setting all else to black (conjunction), copies everything outside of its boundary, setting all else to white (disjunction), copies everything outside of its boundary while inverting all interior samples (exclusive or), or inverts the single source image (negation).

These logical operators can also be extended to grayscale and color images. Corresponding bits of the two source samples serve as logical values where a 1 bit represents true and a 0 bit represents false. Each of the two sources must use the same color model, color depth and dimension. The logical operations are rarely used in color image processing but are nonetheless useful in certain applications.

5.8.1   Implementing the Logical Operations

The logical image operations are implemented by writing concrete subclasses of the BinaryImageOp as was done with the arithmetic operations. Listing 5.12 is a complete implementation of disjunction. The combine method uses the bitwise-or operator, which is denoted by the vertical bar of line 7. This image operation will work correctly for color, grayscale, and binary images.

```
1  public class OrBinaryOp extends BinaryImageOp {
2    public OrBinaryOp(BufferedImage left){
3      super(left);
4    }
5
6    public int combine(int s1, int s2) {
7      return s1 | s2;
8    }
9  }
```

Listing 5.12. OrBinaryOp.

5.9   Alpha Blending

Alpha blending is closely related to the notion of image arithmetic. In alpha blending two sources are added together after being proportionally scaled with an α value. Alpha blending is defined as

$$s_{\text{dst}} = \alpha \cdot s_1 + (1 - \alpha) \cdot s_2,$$

where α is a scalar value between 0 and 1. When $\alpha = 0$ the output image is identical to I_b since I_a is weighted to zero. When $\alpha = 1$ the output image is identical to I_a since I_b is weighted to zero. When $\alpha = .5$ the output image is the average of the two sources and is equivalent to image addition with rescaling. Alpha blending supports the notion of transparency where an alpha value of zero corresponds to complete transparency and a value of 1 corresponds to a completely opaque image.

While alpha blending allows an entire image to have a transparency setting, alpha compositing is a more sophisticated process that allows image processing and computer graphics pipelines to support transparency at the pixel level. Alpha compositing is heavily used in the computer graphics rendering pipeline to combine multiple layers into a single frame. In Java, the BufferedImage class supports an RGBA color model, where the fourth channel is an α channel that determines the per-pixel transparency values.

5.10   Other Blending Modes

The binary image operations presented in this chapter include the arithmetic and logical operations in addition to alpha blending. Other blending modes are in common use for more specialized effects.

The *darken only* blending mode returns the minimum of the two source samples while the *lighten only* mode returns the maximum of the two source samples.

Diffusion is where two samples are combined by selecting one of the two samples at random as the result. It is somewhat inaccurate to describe this as a combination of samples since one of the samples is always ignored and contributes nothing to the result. If the first source sample has an alpha channel, that alpha value can be taken as the likelihood of taking that sample over the second source sample. If, for example, the alpha value of s_1 is 1 then that sample will always be taken over s_2, whereas if s_1 has an alpha of .5 then there is a 50% chance of taking either sample. In the absence of an alpha channel the client can be allowed to select the global odds of taking one source over the other.

Screen blending is defined in Equation (5.15), where the samples are assumed to be normalized. In this mode, the two source images are inverted, then multiplied and finally inverted again to produce the result. Screen blending tends to increase the brightness of an image such that dark areas are brightened more than light areas. Screen blending an image with itself is an effective way of correcting an underexposed image and increasing the detail of dark areas of an image.

$$s_{\mathrm{dst}} = 1 - ((1 - s_1) \cdot (1 - s_2)). \tag{5.15}$$

Equation (5.16) defines *hard light* blending where the image samples are again assumed to be normalized. Hard light blending lightens or darkens an image depending on the brightness of the rightmost image. Blend colors in the right image that are more than 50% brightness will lighten the base image in the same way as the screen blending mode. Blend colors that are less than 50% brightness in the right image will darken the image in the same way as the multiply blending mode. If the rightmost image is all black, the result will be black, if the rightmost image is all white, the result will be white, and if the rightmost image is mid gray, the destination will remain unchanged. Hard light blending is effective for adding shadows to an image or for adding a glow around objects in an image.

$$s_{\mathrm{dst}} = \begin{cases} 2 \cdot s_1 \cdot s_2 & s_2 < .5, \\ 1 - 2 \cdot (1 - s_1) \cdot (1 - s_2)) & s_2 \geq .5. \end{cases} \tag{5.16}$$

5.11   Exercises

1. Since human perception logarithmically compresses light, brightness adjustment is sometimes expressed as a variant of the rescaling technique

described in this chapter. Write a class named `BrightnessAdjustOp` that applies the following relationship between a source sample S and a destination sample S'. In this formulation, we can use a base 10 logarithm and understand that β controls the brightness adjustment:

$$\log S' = \log S + \beta.$$

2. Write a class named `CustomRescaleOp` that implements `BufferedImageOp` and duplicates the behavior of the native `RescaleOp`. The class must use lookup tables. Make sure to include a test routine that applies a `RescaleOp` and a `CustomRescaleOp` to a source image and ensures that the resulting images are identical. You should test your class on a variety of images including grayscale, binary, and true color images.

3. Write a class named `SlowRescaleOp` that implements `BufferedImageOp` and duplicates the behavior of the native `RescaleOp` but does *not* use lookup tables. Apply this filter to some source images and compare the length of time the `SlowRescaleOp` filter takes to complete with the length of time taken by the `CustomRescaleOp` of the previous problem.

⋆ 4. Write a class named `BandExtractOp` that implements `BufferedImageOp`. The filter method must take a source image and return one band of the source as a grayscale image. When the op is constructed, the band to extract must be specified. The `BandExtractOp` must allow the following bands to be extracted from any source image regardless of the color model of the source: R, G, B, C, M, Y, K, Y, I Q, H, S, B.

5. Write a program that applies pseudo coloring to a grayscale image. The program must allow users to

   (a) select an image file,

   (b) view the grayscale source image,

   (c) select a palette file,

   (d) and apply the selected palette to the grayscale score and display the result.

   The palette file contains exactly 256 lines of text. The ith line of the file contains three whitespace-separated integer numbers that represent the RGB values of the ith color of the palette.

6. Write a program to perform histogram equalization on a user-selected image and display the result.

7. Write a program to characterize a grayscale image as dark, light, or gray and poor contrast or good contrast. An image should be designated as *dark* if 80% of its pixels fall below the threshold of 100. An image should be designated as *light* if 80% of its pixels fall above the threshold of 155 and an image should otherwise be designated as *gray*. In addition, an image occupies less than 65% of the dynamic range then the image is characterized as poor contrast while an image that covers at least 90% of the dynamic range is characterized as good contrast.

8. Write a program to perform local histogram equalization on a user-selected image and display the result. Local histogram equalization is achieved by

   (a) selecting a pixel p,

   (b) computing the cumulative distribution function of the $N \times N$ region centered on p,

   (c) converting the CDF into a lookup table,

   (d) transforming pixel p according to the generated lookup table.

   These steps should be performed for every pixel in the image in order to perform local equalization. When the region extends beyond the image boundaries the output value is simply set to be zero.

9. Write concrete subclasses of the `BinaryImageOp` that perform image multiplication, subtraction, and division.

Artwork

Figure 5.4. *Vincent's Bedroom in Arles* by Vincent Willem van Gogh (1853–1890). Vincent van Gogh was a Dutch Impressionist painter who has since become one of the most popular artists of all time. Van Gogh is well known for his use of strong, vivid colors and unique undulating style. He was a prolific artist who produced thousands of pieces during his life. The image of Figure 5.4 is one of five works depicting this scene and is among the best of his works in terms of style and maturity. In a letter to his brother, van Gogh writes of this work that "My eyes are still tired by then I had a new idea in my head [sic] and here is the sketch of it. This time it's just simply my bedroom, only here color is to do everything, and giving by its simplification a grander style to things, is to be suggestive here of rest or of sleep in general. In a word, looking at the picture ought to rest the brain, or rather the imagination. And that is all—there is nothing in this room with its closed shutters. I shall work on it again all day, but you see how simple the conception is. The shadows and the cast shadows are suppressed; it is painted in free flat tints like the Japanese prints."

Figure 5.5. *Portrait of Dr. Gachet* by Vincent Willem van Gogh (1853–1890). See previous entry for a brief biography of the artist. Dr. Paul Gachet was a physician who cared for van Gogh during his last months of life. Van Gogh painted two distinct portraits of Dr. Gachet which both show him sitting at a table leaning on his right arm. The surrounding scening elements, however, are visibly different and hence it is easy to distinguish between the two versions. While the portrait of Figure 5.5 was, at the time of this writing, owned by the Musée d'Orsay, the other version sold in 1990 for US$82.5 million, which, at the time, made it the most expensive painting in the world.

Figure 5.7. *The Girl with a Pearl Earring* by Johannes Vermeer (1632–1675). Johannes Vermeer or Jan Vermeer was a Dutch Baroque painter of the 17th century. He lived his entire life in the town of Delft from which his inspiration for his portraits of ordinary domestic life was drawn. Although Vermeer was not well known in his time, he is today considered to be one of the greatest painters of the Dutch Golden Age. Figure 5.7 is one of his masterworks and is sometimes referred to as "the Mona Lisa of the North" or "the Dutch Mona Lisa."

Figure 5.8. "Ivey Mill" by Lewis Hine. See page 50 for a brief biography of the artist. Figure 5.8 is an image of young Ivey Mill. Hine's full title is "Ivey Mill. Little one, 3 years old, who visits and plays in the mill. Daughter of the overseer. Hickory, N.C." The image is available from the Library of Congress Prints and Photographs Division under the digital ID cph 3a48537.

Figure 5.10. "George Sand" by Gaspard-Felix Tournachon (1820–1910). Gaspard-Felix Tournachon was born in Paris and during his professional life went by the pseudonym of Nadar. He was a caricaturist who discovered portrait photography and was instrumental in the developing years of impressionism. He photographed many of the emerging artists and writers of his time including the likes of Charles Baudelaire, Camille Corot, Eugene Delacroix, Franz Liszt, Victor Hugo, and, as shown in Figure 5.10, George Sand. Please note that the image has been artificially underexposed to suite the purposes of this text and that the original image taken by Tournachon possesses no such defect.

Figure 5.11. "Breaker Boys in #9 Breaker" by Lewis Hine (1874–1940). See page 50 for a brief biography. Figure 5.11 is fully titled "Breaker Boys in #9 Breaker, Hughestown Borough, PA. Coal Co. Smallest boy is Angelo Ross." The image is available from the Library of Congress Prints and Photographs Division under the digital ID cph 3a24720.

Figure 5.12. *Portrait of Jeanne Sammary* by Pierre-Auguste Renoir (1841–1919). Renoir was a French Impressionist painter. In 1862 he began studying art in Paris, where he befriended Alfred Sisley, Frédéric Bazille, and Claude Monet.

Renoir lived in near poverty during the early part of his career due in part to the upheaval of the Fanco-Prussian War. Renoir's paintings use soft but vibrant light and highly saturated color to capture casual snapshots of daily life. The source image of Figure 5.12 is representative of his use of color and his impressionistic style.

Figure 5.13(a). *The Sacrifice of Isaac* by Michelangelo Merisi da Caravaggio (1571–1610). Caravaggio was born in Milan, Italy and was exposed at an early age to some of history's finest works of art, including Leonardo da Vinci's *Last Supper*. Caravaggio led an extremely tumultuous life and, as a youth, fled Milan to Rome due to ongoing conflicts with acquaintances and brawls with the police. In Rome, he secured a series of valuable commissions and became one of the most famous painters in all of Europe. While his style is extremely realistic and emotive it was his tendency towards realism that brought controversy. While most artists of his time depicted Biblical characters and secular heros in idealized form, Caravaggio depicted them as real men with all the ensuing imperfections of body. Figure 5.13 is illustrative of his style as the artist depicts Abraham as an aged and balding man.

Figure 5.13(b). *The Appearance of Christ to the People* by Alexander Andreyevich Ivanov (1806–1858). Ivanov was a Russian painter of the Neoclassical tradition who was raised in St. Petersburg but spent most of his adult life in Rome. His subject matter focused largely on Biblical themes as the detail of Figure 5.13 reflects. Ivanov labored for 20 years on his solitary masterpiece entitled *The Appearance of Christ to the People* which measures an impressive 18×25 feet in dimension. The painting depicts John the Baptist announcing the arrival of Jesus in fulfillment of Old Testament prophecy.

Figure 5.14. "Salvador Dali" by Carl Van Vechten (1880–1964). Carl Van Vechten was born in Cedar Rapids, Iowa, where he developed a keen interest in music and theater, which eventually drew him to attend the University of Chicago to study the arts. He was introduced to photography in the early 1930s and became a prolific portrait photographer. Some of his well-known subjects include F. Scott Fitzgerald, Langston Hughes, Alfred A. Knopf, Bessie Smith, Gertrude Stein, and Salvador Dali, who is shown in Figure 5.14. The Dali portrait is available from the Library of Congress Prints and Photographs Division under the digital ID cph 3c16608.

Figures 5.15 and 5.16. *Portrait of Lunia Czechowska* by Amedo Modigliani (1884–1920). Amedo Modigliani was an Italian painter and sculptor of Jewish heritage. Modigliani's style is unique and difficult to categorize since it draws on elements of many genres and art movements. Modigliani was a prolific artist whose intensely energetic pace and overuse of alcohol likely contributed to his early death at the age of 35. His portrait as shown in Figures 5.15 and 5.16 show his unique style and exhibit the influences of both primitive and impressionistic styles.

Figure 5.17. *The Fall of Satan* by Gustave Doré (1832–1883). Gustave Doré was born in Strasbourg and is well known for his work as a literary illustrator working primarily as an engraver. He is best known for his illustrations of the English Bible but also produced excellent illustrations for *Don Quixote*, "The Raven," "Rime of the Ancient Mariner," *The Divine Comedy*, and, as shown in Figure 5.17, *Paradise Lost*.

Regional Processing Techniques

6

6.1 Overview

Two fundamentally distinct types of image enhancement techniques exist: those that operate within the spatial domain and those that operate in the frequency domain. While these terms are more clearly delineated in Chapter 9 we informally define the spatial domain as a way of representing an image by spatially arranging pixels (colors) to form a complete whole. The frequency domain is an alternate way of representing visual information by describing patterns of global fluctuation in images. This chapter focuses exclusively on spatial domain techniques, leaving the more difficult frequency domain issues for Chapter 9.

Spatial domain techniques can be further classified as either point or regional processing. This classification scheme was given in Chapter 5 and is repeated in Figure 6.1 for the reader's convenience. This chapter focuses exclusively on spatial-domain regional processing techniques.

Under point processing a single input sample is processed to produce a single output sample. In regional processing the value of the output sample is dependent on the values of samples within close proximity to the input sample. Exactly how the region is selected and how samples within the region affect the output depends upon the desired effect.

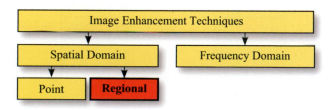

Figure 6.1. Taxonomy of image processing techniques.

6.2   Convolution

The most important regional processing technique is known as convolution. Convolution establishes a strong connection between the spatial and frequency domains and can be used to achieve a wide variety of effects. Convolution employs a rectangular grid of coefficients, known as a kernel, to determine which elements of an image are in the neighborhood of a particular pixel; and the kernel also dictates how these neighborhood elements affect the result.

Figure 6.2 gives an example of a 3×3 kernel, where the center of the kernel, also known as the *key element* is designated as element $K(0,0)$. For a kernel to possess a center element, both the width and height of a kernel must be of odd dimension. While the most frequently used kernels have odd dimension, there are a handful of kernels in popular use that are of even dimension. Each of the kernel elements is a real-valued number and may be either positive or negative. The size of the kernel corresponds to the size of the region while the values of the kernel dictate the effect that regional samples have on the destination.

Figure 6.2 shows an image I where a single sample, $I(x, y)$, is processed by convolution with K. Since the kernel dimension is 3×3, those pixels that affect the result are located in the 3×3 rectangular region centered on $I(x, y)$. Convolving $I(x, y)$ with K produces an output sample $I'(x, y)$, which is obtained by the process illustrated in Figure 6.3. The output sample $I(x, y)$ is obtained by reflecting the kernel about the key element in both the horizontal and vertical axes; placing the kernel over the source sample $I(x, y)$ such that the key element overlays $I(x, y)$; multiplying each of the kernel coefficients by the corresponding source sample; and finally summing these nine products to obtain the output sample $I'(x, y)$.

Each output sample of convolution is a weighted combination of the samples in the surrounding neighborhood. Convolution is a linear process since addition and multiplication are the only arithmetic operations involved when combining

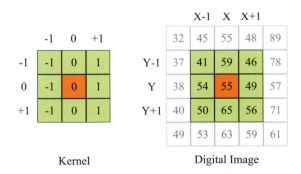

Figure 6.2. A 3×3 kernel is centered over sample $I(x, y)$.

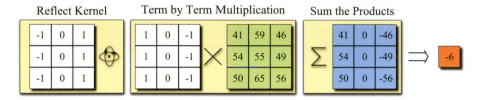

Figure 6.3. Convolution steps.

regional samples. While Figure 6.3 gives an informal description of convolution, we now provide a formal definition. Given a $W \times H$ grayscale image I and an $M \times N$ kernel K where both M and N are odd, the convolution of I with K is denoted as $I \otimes K$ and is defined in Equation (6.1), where $x \in [0, W - 1]$ and $y \in [0, H - 1]$:

$$(I \otimes K)(x, y) = I'(x, y) = \sum_{j=-\lfloor M/2 \rfloor}^{\lfloor M/2 \rfloor} \sum_{k=-\lfloor N/2 \rfloor}^{\lfloor N/2 \rfloor} K(j, k) \cdot I(x - j, y - k). \quad (6.1)$$

The kernel is not literally rotated by copying data from one element to another within some two dimensional data structure but is achieved through careful definition of index values. Specifically, the term $I(x - j, y - k)$ effectively rotates the kernel by multiplying a kernel element $K(j, k)$ with the source element at $I(x - j, y - k)$ rather than the source element at $I(x + j, y + k)$. When Equation (6.1) is applied to the kernel and image of Figure 6.2 the output sample is obtained as

$$\begin{aligned} I'(x, y) &= K(-1, -1) \cdot I(x + 1, y + 1) + K(+0, -1) \cdot I(x, y + 1) + K(1, -1) \cdot I(x - 1, y + 1) \\ &+ K(-1, +0) \cdot I(x + 1, y + 0) + K(+0, +0) \cdot I(x, y + 0) + K(1, +0) \cdot I(x - 1, y + 0) \\ &+ K(-1, +1) \cdot I(x + 1, y - 1) + K(+0, +1) \cdot I(x, y - 1) + K(1, +1) \cdot I(x - 1, y - 1). \end{aligned}$$

Substituting actual coefficients and samples into the above relation yields

$$\begin{aligned} I'(x, y) &= -1 \cdot 56 + 0 \cdot 65 + 1 \cdot 50 \\ &+ -1 \cdot 49 + 0 \cdot 55 + 1 \cdot 54 \\ &+ -1 \cdot 46 + 0 \cdot 59 + 1 \cdot 41 \\ &= -6. \end{aligned}$$

6.2.1 Computation

Equation (6.1) can be directly translated into Java code in a naive fashion. The kernel can be represented as a two-dimensional array of floats, the summations are converted into for loops, the formula of Equation (6.1) is applied to every sample $I(x, y)$, and careful handling of the kernel indices allows us to implicitly set the kernel center to be at location $K(0,0)$. Listing 6.1 gives the straightforward and naive implementation of convolution where the kernel is represented as a two-dimensional array of floats and is assumed to be accessible within the getSumOfTerms method.

```java
public class SimpleConvolveOp extends NullOp {
  private int width,  height;
  private float[][] kernel;

  public SimpleConvolveOp(int width, int height, float[][] k) {
    kernel = k;
    this.width = width;
    this.height = height;
  }

  private double convolvePoint(BufferedImage src, Location pt) {
    int mLow = width / 2;
    int nLow = height / 2;
    int mHigh = width % 2 == 1 ? mLow : width - mLow - 1;
    int nHigh = height % 2 == 1 ? nLow : height - nLow - 1;

    float sum = 0;
    for (int k = -nLow; k <= nHigh; k++) {
      for (int j = -mLow; j <= mHigh; j++) {
        try {
          sum += kernel[k + nLow][j + mLow] *
                 src.getRaster().getSample(pt.col - j, pt.row - k, pt.band);
        } catch (ArrayIndexOutOfBoundsException e) {
        }
      }
    }

    return sum;
  }

  public BufferedImage filter(BufferedImage src, BufferedImage dest) {
    if (dest == null) {
      dest = createCompatibleDestImage(src, src.getColorModel());
    }
    for (Location pt : new RasterScanner(src, true)) {
      int output = ColorUtilities.clamp(convolvePoint(src, pt));
      dest.getRaster().setSample(pt.col, pt.row, pt.band, output);
    }
    return dest;
  }
}
```

Listing 6.1. SimpleConvolutionOp.

In this implementation, the `filter` method iterates over every source sample and obtains the point-wise convolution from the `convolvePoint` method. Within the `convolvePoint` method, the kernel indices are computed such that the kernel indices range over [nLow, nHigh] in height and [mLow, mHigh] in width. Properly computing these boundaries is dependent upon whether the kernel dimensions are odd or even, as shown in lines 14 and 15. In addition, the `convolvePoint` method is problematic at the image boundaries since the `getSample` method of line 22 will generate invalid indices if the column or row index is smaller than the corresponding kernel dimension. While the implementation of Listing 6.1 is correct, it is inefficient and inflexible for practical use. The issues of flexibility and efficiency are discussed in the following sections.

Edge handling. Two important implementation issues must be addressed when performing convolution. The first relates to how the edges of a source image are processed, and the second concerns the color depth of the output. Edge handling is problematic since, when centering a kernel over every element of a source image, the kernel will extend beyond the source image boundaries near the image edge. This situation was noted in line 22 of Listing 6.1 and is visually illustrated in part (a) of Figure 6.4, where a 3×3 kernel is centered on sample $I(0,0)$ of a 6×4 source image. Since there are no source samples corresponding to the top row and left edge of the kernel, Equation (6.1) is ill-defined. Kernels larger than the source image present an extreme situation where convolution is ill-defined for all of the source samples since the kernel extends beyond the image boundaries wherever it is placed. This situation is illustrated in part (b) of Figure 6.4, where a 5×5 kernel is convolved with a 5×2 source image.

The edge problem exhibits itself as an `ArrayOutOfBounds` exception in Listing 6.1. Consider, for example, the `getSample` method nested within the interior `for` loop on line 22. When $x = 0$, $y = 0$, $j = 1$, and $k = 1$ `getSample` is called as `getSample(-1,-1,0)` and throws an exception since the image indices $(-1, -1)$ are invalid.

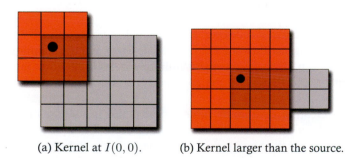

(a) Kernel at $I(0,0)$.   (b) Kernel larger than the source.

Figure 6.4. Illustration of the edge handling problem.

Solutions to the edge problem work by either (1) eliminating edges from the image or (2) redefining convolution at boundary samples. Although it may seem difficult to eliminate edges from an image, this technique is the most common and is generally simple to implement. In order to remove edges we must treat the source image as if it were infinitely wide and tall. This is done through some type of image padding. The following list summarizes the most common approaches to solving the edge handling problem.

♦ *Convolution is redefined to produce zero when the kernel falls off of the boundary.* If the kernel extends beyond the source image when centered on a sample $I(x, y)$ then the output sample is set to zero. This effectively ignores the boundary pixels and creates a black border around the convolved image.

♦ *Convolution is redefined to produce $I(x, y)$ when the kernel falls off the boundary.* If the kernel extends beyond the source image when centered on a sample $I(x, y)$ then the output sample is defined as $I(x, y)$. This effectively copies the input to the convolved output at the boundaries.

♦ *Extend the source image with a color border.* The source image is infinitely extended in all directions by adding zero-valued samples so that samples exist *off the edge* of the source. This is often referred to as zero padding. Since the source image is infinite in extent, convolution is well defined at all points within the bounds of the unpadded source.

♦ *Extend the source image by circular indexing.* The source image is infinitely extended in all directions by tiling the source image so that samples exist *off the edge* of the source. Again, since the source image is infinite in extent, convolution is well defined at all points within the bounds of the unpadded source. Tiling is achieved through the use of circular indexing.

♦ *Extend the source image by reflective indexing.* The source image is infinitely extended in all directions by mirroring the source image so that samples exist *off the edge* of the source. This mirroring is achieved through the use of reflective indexing.

Zero padding, circular indexing, and reflected indexing effectively create an infinitely large image as illustrated in Figure 6.5. Each of these techniques should be understood to pad the source image infinitely in both the horizontal and vertical dimension and thus solve the edge problem since there is no longer any edge.

Of course creating an infinitely large data structure is not possible, so padding must be done by writing methods that produce correct sample values if the indices extend beyond the boundaries of the image that is to be padded. This can be achieved by intercepting the getSample method of the Raster class and augmenting it with a customized getSample method.

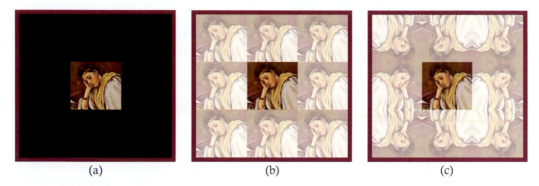

Figure 6.5. (a) Zero padding, (b) circular indexing, and (c) reflected indexing.

We will create an `ImagePadder` interface that represents an image that is infinite in extent. The interface will contain a single method that will accept a source image and a set of indices and return the sample at the specified location. The `ImagePadder` interface is placed in the `pixeljelly.utilities` package and is shown in Listing 6.2.

```
1 public interface ImagePadder {
2   public int getSample(BufferedImage src, int col, int row, int band);
3 }
```

Listing 6.2. `ImagePadder`.

We will now create three `ImagePadder` classes that correspond to zero padding, tiling, and circular indexing. Listing 6.3 gives a complete implementation of an `ImagePadder` for zero padding an image. The `getSample` simply

```
1 public class ZeroPadder implements ImagePadder {
2   public int getSample(BufferedImage src, int col, int row, int band) {
3     if(col < 0 || col >= src.getWidth() || row < 0 || row >= src.getHeight()) {
4       return 0;
5     } else {
6       return src.getRaster().getSample(col, row, band);
7     }
8   }
9 }
```

Listing 6.3. `ZeroPadder`.

checks the row and column indices for validity and if the indices extend beyond the image boundaries the value 0 is returned. If the row and column indices are valid, the corresponding sample in the source image is returned.

We will next develop a class for tiling an image. When tiling an image we need to use circular indexing such that the getSample method will treat all indices as though they are repeating sequences that wrap around at the boundaries of the source image. If, for example, sample $I(W, 0)$ is accessed in a $W \times H$ source image, the sample corresponding to that location will be found at $I(0, 0)$ of the source. Implementation is straightforward since the remainder operator can be employed to correctly transform all indices into valid locations on the source image. Listing 6.4 gives a complete solution for circular indexing; where note care must be taken to properly handle negative index values, as shown in lines 6 and 9.

```
1  public class TiledPadder implements ImagePadder {
2    public int getSample(BufferedImage src, int col, int row, int band) {
3      col = col % raster.getWidth();
4      row = row % raster.getHeight();
5      if (col < 0) {
6        col += raster.getWidth();
7      }
8      if (row < 0) {
9        row += raster.getHeight();
10     }
11     return src.getRaster().getSample(col, row, band);
12   }
13 }
```

Listing 6.4. `TiledPadder`.

When using reflective indexing, the customized code for accessing a sample should treat all indices as though they are reflected back onto the source image. The source image is essentially treated as though surrounded by mirrors on all four sides and hence if an index wanders off of the edge, it is reflected back onto a valid sample of the source image. If, for example, sample $I(W, 0)$ is accessed in a $W \times H$ source image, the sample corresponding to that location will be found at $I(W - 1, 0)$ of the source. Listing 6.5 gives an implementation of reflective indexing. Most textbook implementations either give incorrect solutions for reflective indexing or place constraints on the index values for which their solution will function correctly. The implementation in Listing 6.5 is a general and constraint-free solution for reflection.

```java
public class ReflectivePadder implements ImagePadder {
    private static ReflectivePadder singleton = new ReflectivePadder();
    private ReflectivePadder() { }

    public static ReflectivePadder getInstance() {
        return singleton;
    }

    public int getSample(BufferedImage src, int col, int row, int band) {
        if (col < 0) col = -1 - col;
        if ((col / src.getWidth()) % 2 == 0) {
            col %= src.getWidth();
        } else {
            col = src.getWidth() - 1 - col % src.getWidth();
        }

        if (row < 0) row = -1 - row;
        if ((row / src.getHeight()) % 2 == 0) {
            row %= src.getHeight();
        } else {
            row = src.getHeight() - 1 - row % src.getHeight();
        }

        return src.getRaster().getSample(col, row, band);
    }
}
```

Listing 6.5. `ReflectivePadder`.

From a software engineering perspective we note that each of the three
ImagePadder classes above is a stateless object consisting of a single method.
In addition, the source image is not modified throughout the body of the code
(i.e., it is a read-only parameter) and hence is intrinsically thread safe. These ob-
servations lead to the conclusion that a single instance of any of these classes is
sufficient for any application. Towards that end we will refactor the code to en-
force the notion that only a single instance is ever created in an application and
that instance is made available through a static accessor. This refactored design
is shown in Listing 6.6.

The singleton pattern is implemented by making the default constructor pri-
vate so that external classes are prevented from constructing a ZeroPadder. A
ZeroPadder object is created as a class-level member, as shown in line 2 and the
getInstance class method allows access to that singleton. The ImagePadder
classes provide a flexible and elegant solution to the border handling problem
and will be used to create an exceptionally adaptable convolution operation that
is presented later in this section.

```
1  public class ZeroPadder implements ImagePadder {
2      private static ZeroPadder singleton = new ZeroPadder();
3      private ZeroPadder() { }
4
5      public static ZeroPadder getInstance() {
6          return singleton;
7      }
8
9      public int getSample(BufferedImage src, int col, int row, int band) {
10         if (col < 0 || col >= src.getWidth() ||
11             row < 0 || row >= src.getHeight()) {
12             return 0;
13         } else {
14             return src.getRaster().getSample(col, row, band);
15         }
16     }
17 }
```

Listing 6.6. ZeroPadder as singleton.

Transfer type. The second issue relating to convolution deals with the transfer type and color depth of the resulting image. Since kernel weights are real values[1] and since there are no limits on the values of the weights, both the type and color depth of the output image are determined by the kernel and may be vastly different than the source.

Consider, for example, an 8-bit grayscale source image that is convolved with a 3×3 kernel where all coefficients have a value of $1/4$ or $.25$. Each output sample will be real valued since multiplication of an integer (the 8-bit input sample) and a real (the kernel coefficient) produces a real (the output sample). In addition, each output sample will fall within the range $[0, 573.75]$ since when the kernel is convolved with an all-white region of the source, the output is $(9 \times 255)/4$. In this example the transfer type of the output image is float and the color depth is on the order of 32 bits.

Convolving a source image I having an int transfer type and 8-bit color depth with a real-valued kernel K produces an output image I' having a float transfer type and a 32-bit color depth. This is problematic since BufferedImage does not fully support the use of floats as a transfer type.

One solution to the color depth problem is to rescale each sample $I'(x, y)$ to the same color depth as the source as convolution is performed. Another solution is to scale the *kernel* such that all processed samples will fall within the desired color depth. When a kernel contains all non-negative coefficients it can be normalized by scaling the kernel such that the coefficients sum to 1. Any

---

[1]In mathematical terms we write that $K(x, y) \in \Re \forall x, y$.

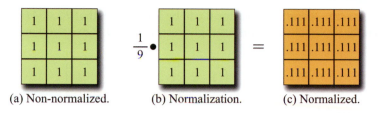

(a) Non-normalized.　　(b) Normalization.　　(c) Normalized.

Figure 6.6. Normalizing a kernel.

kernel having all nonnegative coefficients can be easily normalized by dividing each kernel coefficient by the sum of the non-normalized coefficients.

Consider the non-normalized 3×3 kernel of Figure 6.6. The maximum output that this kernel can produce occurs when all samples in a 3×3 region of the source image are maximal. For an 8-bit image the maximum possible value will by $255 \times 9 = 2,541$. The kernel can be prescaled, however, by dividing each coefficient by the sum of all coefficients, as shown in (b). The normalized kernel will produce a maximum output sample of $255 \times .111 = 255$, thus preserving the color depth.

An alternate solution to rescaling the kernel is to clamp the output to 8-bits regardless of the kernel coefficients. This tactic is often adopted for image processing applications since it maintains the color depth (8 bits) and transfer type (integer values) of the source image. The responsibility of providing well-tuned kernels then falls on the convolution clients.

6.2.2 Analysis

Convolution is a computationally expensive operation when implemented in straightforward fashion. Consider a $W \times H$ single band image and a $M \times N$ kernel. The source image contains $W \times H$ samples, each of which must be convolved with the kernel. Since a total of $M \times N$ multiplications and $M \times N$ additions are required to compute the output for a single sample, the total number of arithmetic operations required for convolution is on the order of $W \times H \times M \times N$. The computational effort is therefore directly proportional to the size of the kernel and hence large kernels should be used sparingly.

Computational efficiency can be dramatically improved if the kernel is *separable*. A $M \times N$ kernel is separable if a $M \times 1$ column vector A and a $1 \times N$ row vector B can be found such that $K = AB$. Consider, for example, the 3×3 kernel given as

$$K = \begin{bmatrix} 1 & 2 & 1 \\ 2 & 4 & 2 \\ 1 & 2 & 1 \end{bmatrix}.$$

If a 3×1 vector A and 1×3 vector B can be found such that $K = AB$ then K is shown to be separable. For this example vectors A and B exist and are given by

$$AB = \begin{bmatrix} 1 \\ 2 \\ 1 \end{bmatrix} \times \begin{bmatrix} 1 & 2 & 1 \end{bmatrix} = \begin{bmatrix} 1 & 2 & 1 \\ 2 & 4 & 2 \\ 1 & 2 & 1 \end{bmatrix}.$$

The benefit of separable kernels is obtained by realizing that convolution can be performed by consecutive application of two separate convolutions using the smaller vectors as kernels. Convolution with a separable kernel is formally given in Equation (6.6).

$$I \otimes K = (I \otimes A) \otimes B \text{ where } K = AB. \tag{6.6}$$

Recall that approximately $W \times H \times M \times N$ arithmetic operations are required to perform convolution on a $W \times H$ image and a nonseparable $M \times N$ kernel. A 3×3 kernel K then requires approximately $9 \times W \times H$ operations. If, however, the kernel is known to be separable, the computational complexity is significantly reduced. Since A is a 3×1 kernel, the first convolution requires $3 \times W \times H$ arithmetic operations and since B is a 1×3 kernel, the second convolution requires $3 \times W \times H$ additional operations for a total of $6 \times W \times H$ arithmetic operations. The total number of arithmetic operations required for convolving an image with a separable kernel is on the order of $(M + N) \times W \times H$. If should be apparent from this formulation that the speedup gained through separability dramatically increases with larger kernel size.

6.2.3   ConvolveOp

Java directly supports convolution through two specific classes. `Kernel` is a class representing a matrix of floating point coefficients. A `Kernel` is constructed by giving the width and height of the kernel along with the coefficients. The coefficients are provided as a single array of floats that are packed in row major order. `ConvolveOp` is a `BufferedImageOp` that is constructed by providing a `Kernel`. Listing 6.7 gives a code fragment that convolves a source image with a kernel of uniform weights.

```
1  float[] weights = {.1f, .1f, .1f, .1f, .1f, .1f, .1f, .1f, .1f};
2  Kernel kernel = new Kernel(3, 3, weights);
3  ConvolveOp op = new ConvolveOp(kernel);
4  BufferedImage source = ImageIO.read(new File("filename.png"));
5  BufferedImage filtered = op.filter(source, null);
```

Listing 6.7. `ConvolveOp` example.

ConvolveOp clients have little control over how the convolution is performed. The ConvolveOp always clamps the destination image into an 8-bit int and does not take advantage of separable kernels. The ConvolveOp handles boundary conditions by either setting the edges to black (zero) or by copying the source image pixels directly. Zero padding, circular indexing, and reflected indexing are not supported.

We now present a customized convolution class structure that is far more robust and flexible than what Java provides. We first develop a set of kernel classes that represent both separable and nonseparable kernels so that convolution can be efficiently performed when the kernel is separable. In order to distinguish it from the Kernel class that is built into Java, we define an abstract kernel class named Kernel2D, as shown in Listing 6.8.

```
1  public abstract class Kernel2D {
2    private Rectangle bounds;
3
4    public Kernel2D(int width, int height) {
5      bounds = new Rectangle(-width/2, -height/2, width, height);
6    }
7
8    public Rectangle getBounds() {
9      return bounds;
10   }
11
12   public int getWidth() {
13     return bounds.width;
14   }
15
16   public int getHeight() {
17     return bounds.height;
18   }
19
20   public abstract boolean isSeparable();
21   public abstract float getValue(int col, int row);
22   public abstract Kernel2D getRowVector();
23   public abstract Kernel2D getColVector();
24 }
```

Listing 6.8. Kernel2D.

The Kernel2D class is an abstract base class where the main purpose is to maintain and provide the bounds of the kernel object. The width and height are stored as a Rectangle in the bounds attribute. The x and y coordinates of the bounds are given in line 5 as $-width/2$ and $-height/2$. Kernel coefficients are accessible through the getValue method; note that the key element is at the center of the kernel and hence the column and row indices are allowed to be negative. If the `Kernel2D` is separable, it is able to provide the row and column vectors as `Kernel2D` objects through the `getRowVector` and `getColVector` methods.

We now author two concrete subclasses of the `Kernel2D` that give meaning to the four abstract methods of the `Kernel2D` class. Listing 6.9 gives the implementation for, as the name obviously implies, a kernel that is nonseparable.

```java
public class NonSeparableKernel extends Kernel2D {
    private float[] values;

    public NonSeparableKernel(int width, int height, float[] values) {
        super(width, height);
        this.values = values;
    }

    @Override
    public float getValue(int col, int row) {
        int index = (row - getBounds().y) * getWidth() + (col - getBounds().x);
        return values[index];
    }

    @Override
    public Kernel2D getRowVector() {
        throw new UnsupportedOperationException("Non seperable kernel.");
    }

    @Override
    public Kernel2D getColVector() {
        throw new UnsupportedOperationException("Non seperable kernel.");
    }

    @Override
    public boolean isSeparable() {
        return false;
    }
}
```

Listing 6.9. `NonSeparableKernel`.

The constructor accepts the coefficients as a linearized array arranged in row major order and having the specified width and height. The `getValue` method first maps the specified indices (which may be negative) into the correct index of the linearized array and returns the requested array element. Since objects of this type are not separable, the `getRowVector` and `getColVector` methods simply throw exceptions when called. Of course, the `isSeparable` method returns false.

Implementation of a `SeparableKernel` class is equally straightforward, as shown in Listing 6.10. In this listing, the constructor accepts two arrays that represent the row vector coefficients and the column vector coefficients. The kernel is understood to have as many columns as the row vector and as many rows as the column vector. These two floating point arrays are then transformed into `NonSeparableKernels` and are provided to clients via the `getRowVector` and `getColVector` methods. Note that the `getValue` method simply throws an exception since it would be computationally inefficient to compute. Development of this method is left as an exercise for the reader.

```java
public class SeparableKernel extends Kernel2D {
    private Kernel2D row, col;

    public SeperableKernel(float[] colVector, float[] rowVector) {
        super(rowVector.length, colVector.length);
        row = new NonSeperableKernel(rowVector.length, 1, rowVector);
        col = new NonSeperableKernel(1, colVector.length, colVector);
    }

    @Override
    public boolean isSeparable() {
        return true;
    }

    @Override
    public float getValue(int col, int row) {
        throw new UnsupportedOperationException("Not supported");
    }

    @Override
    public Kernel2D getRowVector() {
        return row;
    }

    @Override
    public Kernel2D getColVector() {
        return col;
    }
}
```

Listing 6.10. `SeparableKernel`.

Listing 6.11 gives a partial listing of a custom convolution operation. This class allows any type of image padding to be utilized, including customized implementation of the `ImagePadder`. Reflective indexing is established as the default border handling mechanism, as shown in line 3 and is utilized in line 10.

The most important aspect of this implementation involves leveraging separable kernels. When a ConvolutionOp is constructed, a Kernel2D is provided. If that kernel is separable then two separate convolutions are performed using the row and column vectors provided by the separable kernel. Also note that the convolveAtPoint method is made impressively concise through the use of a RasterScanner and solid design of the Kernel2D class.

```java
public class ConvolutionOp extends NullOp {
  private boolean takeAbsoluteValue;
  private ImagePadder padder = ReflectivePadder.getInstance();
  private Kernel2D kernel;

  public float convolveAtPoint(BufferedImage src, Location pt) {
    float sum = 0;
    for (Location kernelPoint : new RasterScanner(kernel.getBounds())) {
      sum += kernel.getValue(kernelPoint.col, kernelPoint.row) *
              padder.getSample(src,
                               pt.col - kernelPoint.col,
                               pt.row - kernelPoint.row,
                               pt.band);
    }
    if(takeAbsoluteValue) sum = Math.abs(sum);
    return ImagingUtilities.clamp(sum, 0, 255);
  }

  public ConvolutionOp(Kernel2D k, boolean f) {
    setKernel(k);
    setTakeAbsoluteValue(f);
  }

  public void setPadder(ImagePadder p) {
    this.padder = p;
  }

  public void setKernel(Kernel2D k) {
    this.kernel = k;
  }

  public void setTakeAbsoluteValue(boolean f) {
    this.takeAbsoluteValue = f;
  }

  public BufferedImage filter(BufferedImage src, BufferedImage dest) {
    if (dest == null) {
      dest = createCompatibleDestImage(src, src.getColorModel());
    }

  if (kernel.isSeparable()) {
      ConvolutionOp firstPass =
        new ConvolutionOp(kernel.getColVector(), takeAbsoluteValue);
```

```
44
45        ConvolutionOp  secondPass =
46          new  ConvolutionOp ( kernel . getRowVector () ,  takeAbsoluteValue );
47        firstPass . setPadder ( padder );
48        secondPass . setPadder ( padder );
49
50        return  secondPass . filter ( firstPass . filter ( src ,  null ),  dest );
51      } else {
52        for  ( Location  pt  :  new  RasterScanner ( src ,  true ))  {
53          dest . getRaster (). setSample ( pt . col ,
54                                          pt . row ,
55                                          pt . band ,
56                                          convolveAtPoint ( src ,  pt ));
57        }
58        return  dest ;
59      }
60    }
61  }
```

Listing 6.11. ConvolutionOp (partial listing).

Note that it is often desirable to take the absolute value of the point convolutions prior to clamping. Our implementation allows for this, as shown in line 15, where the takeAbsoluteValue attribute controls when this feature is enabled.

6.3   Smoothing

Convolution can be used to achieve a variety of effects depending on the kernel. Smoothing, or blurring, can be achieved through convolution and is often used to reduce image noise or to prepare an image for further processing stages. Smoothing is accomplished by any kernel where all of the coefficients are nonnegative. Two important classes of smoothing filters are commonly used.

A uniform filter is based on computing the average of all neighborhood samples. Any kernel where all of the nonzero coefficients are identical is a uniform filter. If a uniform filter contains zero-valued coefficients, these pixels are not part of the averaging computation and exert no influence on the processed output. Uniform filters are typically square and normalized such that the coefficients sum to unity. A normalized uniform filter produces a true average of all neighborhood samples while a non-normalized uniform filter yields a scaled average. Figure 6.7 shows two uniform filters where the box filter is named from the shape of the neighborhood (a boxed region) and the circular box filter is likewise named after the nearly circular shape of its neighborhood.

$$\frac{1}{25} \times \begin{bmatrix} 1 & 1 & 1 & 1 & 1 \\ 1 & 1 & 1 & 1 & 1 \\ 1 & 1 & 1 & 1 & 1 \\ 1 & 1 & 1 & 1 & 1 \\ 1 & 1 & 1 & 1 & 1 \end{bmatrix} \qquad \frac{1}{21} \times \begin{bmatrix} 0 & 1 & 1 & 1 & 0 \\ 1 & 1 & 1 & 1 & 1 \\ 1 & 1 & 1 & 1 & 1 \\ 1 & 1 & 1 & 1 & 1 \\ 0 & 1 & 1 & 1 & 0 \end{bmatrix}$$

(a) Box filter.   (b) Circular box filter.

Figure 6.7. Normalized uniform filters.

A weighted smoothing filter uses coefficients that are larger near the center and smaller near the periphery of the kernel. Unlike the averaging filter, which treats all neighborhood pixels as equal with respect to their effect on the result, a weighted averaging filter gives more importance to those samples that are near the key element and less importance to those samples that are further away from the key element. Weighted averaging filters are based on the premise that samples in close proximity to each other are likely to be similar and that this likelihood decreases with distance. When the kernel weights are viewed as a three-dimensional surface the height of the surface at each location indicates the weight or importance at that kernel location. Weighted blur filters will then typically appear as either a pyramid, cone, or two-dimensional Gaussian, as shown in Figure 6.8.

The function for creating a pyramidal surface is given in Equation (6.7). In this formulation, the height of a point in the pyramid is a function f of location where the origin is at the very top of the pyramid and the slope of the pyramid sides is given by α. The pyramid height *decreases* from zero as the distance from the center of the pyramid *increases*. Since pyramids typically appear *above* ground, the formula of Equation (6.7) can be rescaled by computing the surface within some range of values and then adding an offset such that the minimum surface

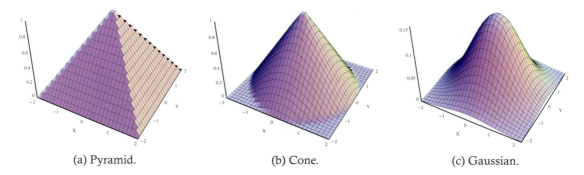

(a) Pyramid.   (b) Cone.   (c) Gaussian.

Figure 6.8. Weighted filters.

1	2	3	2	1
2	4	6	4	2
3	6	9	6	3
2	4	6	4	2
1	2	3	2	1

(a) Pyramid.

0	0	1	0	0
0	2	2	2	0
1	2	5	2	1
0	2	2	2	0
0	0	1	0	0

(b) Cone.

1	4	7	4	1
4	16	28	16	4
7	28	49	28	7
4	16	28	16	4
1	4	7	4	1

(c) Gaussian.

Table 6.1. Discretized kernels.

is at ground level or zero.

$$f(x, y) = -\alpha \cdot \max(|x|, |y|) \tag{6.7}$$

The function for a conical surface is given in Equation (6.8). Again, the height of the cone is a function f of location, where the origin is at the very top point of the cone and the height *decreases* with distance from the center:

$$f(x, y) = -\alpha \cdot \sqrt{x^2 + y^2}. \tag{6.8}$$

The Gaussian of Figure 6.8 has significant advantages over the pyramid and cone. While the pyramid and cone are both separable functions they are not smooth, the pyramid at the four edges and the cone at its peak point. Gaussian blurring is a preferred technique that uses the Gaussian function as a kernel for convolution. The Gaussian kernel is separable, everywhere continuous, and slopes to zero at the kernel boundaries. This latter feature has significant advantages that are discussed in Chapter 9. Gaussian blurring generally provides sufficient smoothing while preserving edges better than the uniform filters or the conic and pyramidal functions. The continuous two-dimensional Gaussian function G is given in Equation (6.9):

$$G(x, y) = \frac{1}{2\pi\sigma^2} e^{\frac{-(x^2 + y^2)}{2\sigma^2}}. \tag{6.9}$$

The surface is smooth and radially symmetric, as shown in part (c) of Figure 6.8. The function peaks at the center and drops off smoothly at the edges thereby giving more weight to samples nearer the center and less weight to samples nearer the edges. The shape of the function, and hence the degree of smoothing, is controlled by the parameter σ which is known as the standard deviation. Larger σ values spread the weights across more surface area and reduce the center peak while smaller σ values form a thinner but taller peak.

The pyramid, cone and Gaussian functions must be imaged in order to be used as convolution kernels. Examples of discretizing the pyramid, cone, and Gaussian function using integer weights are shown in Table 6.1.

A discrete Gaussian kernel is obtained by choosing an appropriate σ value and then imaging the continuous Gaussian function over an appropriate range of coordinates. Larger standard deviations require larger kernel dimensions since the Gaussian is spread over a larger surface.

Since the Gaussian kernel is separable there must exist vectors A and B such that $K = AB$ for all Gaussian kernels K and since the Gaussian kernel is radially symmetric it follows that $A = B$. Generating a Gaussian kernel K can then be done automatically by computing the one-dimensional vector A. Generally speaking, σ determines the amount of smoothing and should be a user-adjustable parameter. The size of the kernel should either be automatically derived from σ or the user should be allowed to also control the size of the kernel independently of the shape of the kernel.

Equation (6.10) is parameterized on both σ and α such that clients are able to independently select both the kernel size and the kernel shape. The standard deviation σ controls the overall shape of the Gaussian and hence the actual values within the vector while the parameter α indirectly controls the size of the kernel that will be generated. In this formulation, W is the half-width of the kernel and is defined as $W = \lceil \alpha\sigma \rceil$. While any value of α can be chosen, values less than 2 tend to truncate the Gaussian function and hence do not provide smoothness at the edges of the kernel; while values of α exceeding 5 include portions of the Gaussian that are insignificant and thus unnecessarily increase the computational load. When Gaussian blurring is supported in commercial image processing applications, the user is generally able to control a *radius* that roughly corresponds to the α value of the Gaussian function:

$$A(n) = \frac{1}{\sigma \cdot \sqrt{2\pi}} e^{\frac{-n^2}{2\sigma^2}} \quad n \in [-W, -W+1, \ldots, W-1, W]. \qquad (6.10)$$

As an example consider selecting $\alpha = 2$ and $\sigma = 1$. The half-width W is computed as $W = \lceil 2 \times 1 \rceil = 2$. The vector A is then known to have five elements corresponding to values of n in $[-2, 2]$. Applying equation (6.10) yields

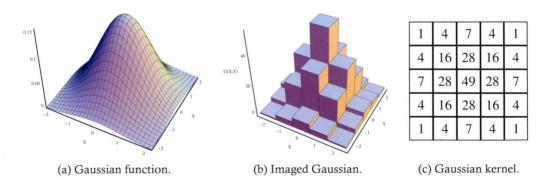

| (a) Gaussian function. | (b) Imaged Gaussian. | (c) Gaussian kernel. |

Figure 6.9. Deriving a Gaussian kernel.

```
1  float[] getGaussianCoefficients(double alpha, double sigma) {
2    int w = (int)Math.ceil(alpha * sigma);
3    float result[] = new float[w*2 + 1];
4    for(int n=0; n<=w; n++){
5      double coefficienet =
6        Math.exp(-(n*n)/(2*sigma*sigma))/(Math.sqrt(2*Math.PI)*sigma);
7      result[w + n ] = (float)coefficient;
8      result[w - n ] = (float)coefficient;
9    }
10   return result;
11 }
```

Listing 6.12. Java method for computing a Gaussian kernel.

the 5 element vector

$$\{A(-2), A(-1), A(0), A(1), A(2)\} = \{0.054, 0.242, 0.399, 0.242, 0.054\}.$$

If integer-valued coefficients are desired, the vector can be rescaled by dividing each term by $\min(A)$ and rounding. For this example, each of the elements is rounded after having been divided by .054 to obtain the approximation given by

$$A = \{1, 4, 7, 4, 1\}.$$

A visual illustration of the derivation of a Gaussian kernel is shown in Figure 6.9, where the surface corresponding to $\sigma = 1$ and $\alpha = 2$ is imaged by sampling and quantizing as shown in (b). The resulting non-normalized 5×5 integer-valued kernel is shown in (c).

(a) Source image.

(b) 17×17 Box.

(c) 17×17 Gaussian.

Figure 6.10. Smoothing examples.

Listing 6.12 gives a Java function for computing the row vector A for use in a separable Gaussian kernel. Note that since the vector is symmetric about its center; only one-half of the values need to be computed but are stored in the array twice as indicated by lines 7 and 8. No attempt is made in this code to normalize the resulting array. While development of a complete class that implements Gaussian filtering is left as an exercise, the class is very simple to implement as a subclass of the previously defined `ConvolutionOp` of Listing 6.11.

Figure 6.10 shows a comparison of box filtering with Gaussian smoothing. Gaussian smoothing, as compared to box filtering, generally provides sufficient blurring while simultaneously preserving important edge details. While not evident from these visual examples, the neighborhood size determines the amount of smoothing performed.

6.4   Edges

An edge is a rapid transition between dark and light areas of an image. The goal of edge detection is to identify locations in an image where such transitions are particularly strong: a large change in light intensity over a short spatial distance. Strong edges are likely to indicate significant visual features of an image such as boundaries around scene elements, changes in texture, and changes in scene depth. While blurring reduces the strength of edges in an image, sharpening is used to increase the strength of edges in an image. Sharpening and edge detection are common techniques in image processing and edge detection is an especially active research area in computer vision.

Edge detection techniques are based on the derivative(s) of an image. The derivative of a one dimensional function $f(x)$ is denoted as $f'(x)$ and is a measure of the amount of change in function f that occurs around location x. Figure 6.11 illustrates how derivatives can be used to identify image edges. The grayscale source image of part (a) has two edges of moderate strength: the leftmost edge rising from black to white while the rightmost edge falls from white to black. A single row of this image is plotted in (b), where the horizontal axis corresponds to the column index and the vertical axis corresponds to the sample's grayscale value. This plot can be understood as a one-dimensional function where light varies with distance from the leftmost column of the image.

The row profile shows a large transition in intensity (black to white) occurring over a moderate spatial distance of approximately 30 pixels (the transition begins near column 40 and levels off near column 70) and thus signals the presence of a rising edge. An identically large transition in intensity (white to black) occurs over a moderate spatial distance of approximately 30 pixels (the transition begins near column 140 and levels off near column 170) and thus signals a descending edge. The derivative is a measure of change at each location of the row plot and

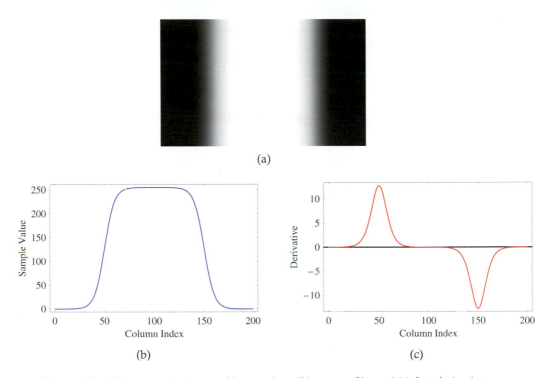

(a)

(b)

(c)

Figure 6.11. (a) A grayscale image with two edges, (b) row profile, and (c) first derivative.

is shown in (c). There is no change at all in the black region at the beginning and end of the row profile nor is there any change in the white central portion of the row profile and hence the derivative is zero at those locations. In the areas of transition, however, the row profile is changing and the change is maximal at the center of both the rising and falling edges. Local extrema (both minimum and maximum values) of the derivative indicate the presence of an edge in the source image and the sign of the derivative indicates whether the edge is ascending or descending.

Images are, of course, functions of two variables rather than one, as shown in the example of Figure 6.11. The derivative of a multi-variate function is known as the gradient and is a measure of the change that occurs in each dimension of the function. The gradient of a continuous 2D function I, denoted by the symbol ∇, is given by Equation (6.13), where $\delta I(x, y)/\delta x$ denotes the derivative of function I with respect to variable x and represents the amount of horizontal change in function I at location (x, y). The second term of the derivate, $\delta I(x, y)/\delta y$, represents the amount of vertical change in I at location (x, y). It is important to note that the gradient at a *single point* consists of two elements. In other words, the gradient at a specific location is a two-dimensional vector, which implies that the gradient of an entire image is a table of such vectors otherwise known as a vector field:

$$\nabla I(x,y) = \left(\begin{array}{c} \delta I(x,y)/\delta x \\ \delta I(x,y)/\delta y \end{array} \right).$$

(6.13)

While the gradient of Equation (6.13) is defined in terms of a Cartesian space, it is often useful to describe the gradient using a polar coordinate system. A two-dimensional vector defined in polar coordinates is a pair (r, Θ), where r is the *radius* or *magnitude* of the vector and Θ is the *orientation* of the vector—the direction in which the vector is pointing. Representation of the gradient in polar coordinates is convenient since the strength and the orientation of the gradient directly convey important information about the content of an image.

Conversion between Cartesian and polar coordinates is straightforward. For conciseness we define $dx = \delta I(x,y)/\delta x$ and $dy = \delta I(x,y)/\delta y$. Given this abbreviation, the gradient can be converted into polar coordinates using Equation (6.14), where the inverse tangent function accepts two arguments and is thus able to discriminate among all four quadrants:

$$\begin{aligned} r &= \sqrt{dx^2 + dy^2}. \\ \Theta &= \tan^{-1}(dy, dx). \end{aligned}$$

(6.14)

In more intuitive terms, the gradient is a measure of the slope, or grade,[2] at all points within the image. Consider a hiker who seeks to climb a hill by taking the most efficient path upwards. When standing at any point on the hill, the gradient tells the hiker in which direction to take his next step (this is the Θ component) and how much higher that step will take him (this is the r, or magnitude, component). The gradient at a particular location is therefore a two-dimensional vector that points toward the direction of steepest ascent where the length (or magnitude) of the vector indicates the amount of change in the direction of steepest ascent. The orientation of the gradient is therefore perpendicular to any edge that may be present.

Figure 6.12 gives a visual example of an image gradient. While the gradient is defined at all points in the image the figure shows, for visual clarity, only a few of the most significant gradient vectors. Each individual vector within the gradient field points in the direction of steepest ascent while the length of the vector indicates the magnitude or grade of the image at that particular point. The most significant vectors are those that indicate the presence of a strong edge or, in other words, those vectors having the largest magnitudes. Since only the strongest gradients are portrayed in this figure they are all about the same strength. The weaker gradients, had they been included, would have been much shorter in length.

---

[2] *Steep Grade Ahead* is a common road sign in the U.S. warning of steep climbs or descents.

Figure 6.12. Image gradient (partial).

6.4.1   Digital Approximation of Gradient

Since digital images are not continuous but discrete, the gradient must be approximated rather than analytically derived. Numerous methods for approximating the gradient of a discrete data set are in common use, but a simple method is illustrated by Figure 6.13. Consider a 3×3 region of an image having samples centered at S_c, where the subscript stands for the *center*, and the surrounding samples are positioned relative to the center as either up/down or left/right as indicated by the subscripts. Given this information, we would like to approximate the amount of change occurring in the horizontal and vertical directions at location (x, y) of image I.

   The amount of change in the horizontal direction can be approximated by the difference between S_r and S_l, while the amount of change in the vertical

	X-1	X	X+1
Y-1	S_{ul}	S_u	S_{ur}
Y	S_l	S_c	S_r
Y+1	S_{dl}	S_d	S_{dr}

Figure 6.13. A 3×3 region centered on sample S_c.

direction can be approximated by the difference between S_d and S_u. In other words, the gradient can be approximated as

$$\nabla I(x,y) = \left(\begin{array}{c} \delta I(x,y)/\delta x \\ \delta I(x,y)/\delta y \end{array} \right) \simeq \left(\begin{array}{c} S_r - S_l \\ S_d - S_u \end{array} \right). \qquad (6.15)$$

Equation (6.15) approximates the gradient at a single location by computing the change in intensity that occurs both horizontally and vertically across the center sample. When the gradient is approximated in this fashion, we make the connection that the gradient of the image can be computed via *convolution*. The kernel corresponding to the horizontal gradient is given by G_x while the vertical kernel is given by G_y in Equation (6.17):

$$G_x = [1, 0, -1] \qquad G_y = \left[\begin{array}{c} 1 \\ 0 \\ -1 \end{array} \right]; \qquad (6.16)$$

such that

$$\nabla I = \left(\begin{array}{c} \delta I/\delta x \\ \delta I/\delta y \end{array} \right) \simeq \left(\begin{array}{c} I \otimes G_x \\ I \otimes G_y \end{array} \right). \qquad (6.17)$$

Figure 6.14 gives a numeric example of how to compute and interpret an image gradient. In this example a grayscale source image (a) has a moderately strong edge extending from the upper left to the lower right of the image. The partial derivatives are computed via convolution by application of Equation (6.17) using the approximations of Equation (6.15) for the horizontal and vertical derivatives. The approximation to the horizontal derivative is given in (b) and the approximation to the vertical derivative is given in (c). The center sample has a gradient of

$$\nabla I(x,y) = \left(\begin{array}{c} 74 \\ -110 \end{array} \right);$$

or a change of 74 units in the horizontal dimension and of -110 units in the vertical dimension. This gradient is shown in (d). The entire gradient field is shown in (e) and the magnitude of the gradient is shown in (f), where it has been rounded to the nearest integer.

When computing the gradient, the transfer type and color depth of the resulting image presents some difficulty. Since the kernel contains negative weights, the output values for both $\delta I/\delta x$ and $\delta I/\delta y$ are able to range over the interval $[-255, 255]$ for an 8-bit source image which implies that the destination has a 9-bit depth. Depending upon the application, the output image is typically either rescaled back into an 8-bit image or the sign of the output image samples is ignored. If the output image is rescaled to 8 bits then strong negative edges will be

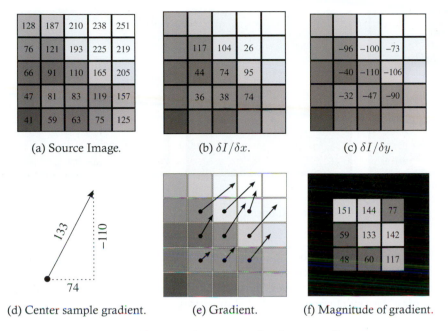

(a) Source Image.   (b) $\delta I/\delta x$.   (c) $\delta I/\delta y$.

(d) Center sample gradient.   (e) Gradient.   (f) Magnitude of gradient.

Figure 6.14. Numeric example of an image gradient.

depicted as black, strong positive edges will be depicted as white, and homogeneous regions will be depicted as middle gray. If the absolute value of the output image is taken, then all strong edges will be depicted as white and homogeneous regions will be depicted as black.

Prewitt operator. The gradient approximation technique of Equation (6.15) is susceptible to image noise since a single noisy or corrupted sample will exert a large influence in the gradient at a particular location. In order to increase the noise immunity, the differences over a 3×3 region can be averaged together. The Prewitt operators are convolution kernels that suppress noise by averaging the differences in the horizontal and vertical directions. The Prewitt kernels are given as

$$G_x = \tfrac{1}{3} \begin{bmatrix} 1 & 0 & -1 \\ 1 & 0 & -1 \\ 1 & 0 & -1 \end{bmatrix}, \qquad G_y = \tfrac{1}{3} \begin{bmatrix} 1 & 1 & 1 \\ 0 & 0 & 0 \\ -1 & -1 & -1 \end{bmatrix}.$$

Figure 6.15 shows an image that is convolved with each of these two kernels and rescaled into an 8-bit image. Black regions show strong negative edges while white regions indicate strong positive edges and mid-gray regions represent homogeneous areas having relatively small change in the source image.

(a) Source image I.

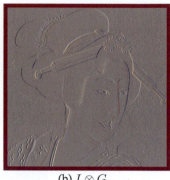
(b) $I \otimes G_x$.

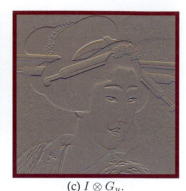
(c) $I \otimes G_y$.

Figure 6.15. Prewitt operators.

Sobel operator. The Sobel operators are similar to the Prewitt operators since they average the differences and hence suppress noise. The Sobel operators improve edge detection, however, by placing greater emphasis on the difference spanning the key element. The Sobel operators are a pair of 3×3 convolution kernels given as

$$G_x = \frac{1}{4} \begin{bmatrix} -1 & 0 & +1 \\ -2 & 0 & +2 \\ -1 & 0 & +1 \end{bmatrix}, \quad G_y = \frac{1}{4} \begin{bmatrix} +1 & +2 & +1 \\ 0 & 0 & 0 \\ -1 & -2 & -1 \end{bmatrix}.$$

Roberts cross gradient operators. The Roberts cross operators are two 2×2 kernels. These operators are different than the Prewitt and Sobel operators since they are sensitive to differences along diagonal axes of the source image rather than the horizontal and vertical axes. The Roberts cross operators are given by

$$G_1 = \begin{bmatrix} +1 & 0 \\ 0 & -1 \end{bmatrix}, \quad G_2 = \begin{bmatrix} 0 & +1 \\ -1 & 0 \end{bmatrix},$$

where the key element is given at the upper-left of each mask.

The Roberts operators, when convolved with an image, measure the amount of change on the diagonals of an image rather than the horizontal and vertical change. Since the Roberts gradient utilizes a smaller kernel, it is more computationally efficient that the Prewitt or Sobel gradients and hence is useful when computational speed is an important factor. Figure 6.16 illustrates the Roberts operators. Note how the diagonal hairpiece in the right central portion of the image is nearly invisible to the G_1 kernel while the G_2 kernel is barely sensitive to the diagonal hairpiece in the left central portion of the source.

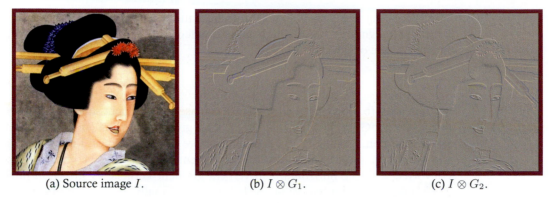

(a) Source image I.   (b) $I \otimes G_1$.   (c) $I \otimes G_2$.

Figure 6.16. A source image is convolved with the Roberts operators G_1 and G_2.

6.4.2 Edge Detection

Edge detection algorithms typically key on the magnitude of the gradient (MoG) since larger magnitudes directly correspond to greater probability of an edge. Computing the MoG is a fundamental building block in image processing. The magnitude of the gradient is denoted as $|\nabla I|$ and has already been defined in Equation (6.14) for a single point within an image. The MoG of an entire image is concisely given by Equation (6.18):

$$|\nabla I| = \sqrt{(I \otimes G_x)^2 + (I \otimes G_y)^2}. \tag{6.18}$$

```
1  public class MagnitudeOfGradientOp extends NullOp {
2    public BufferedImage filter(BufferedImage src, BufferedImage dest) {
3      NonSeperableKernel rowV = NonSeperableKernel(3, 1, new float[]{-1, 0, 1});
4      NonSeperableKernel colV = NonSeperableKernel(1, 3, new float[]{-1, 0, 1});
5
6      ConvolutionOp horizontalOp = new ConvolutionOp(rowV, true);
7      ConvolutionOp verticalOp = new ConvolutionOp(colV, true);
8
9      BufferedImage vGradient = verticalOp.filter(src, null);
10     BufferedImage hGradient = horizontalOp.filter(src, null);
11     return new AddBinaryOp(vGradient).filter(hGradient, null);
12   }
13 }
```

Listing 6.13. MagnitudeOfGradientOp.

Calculating the square root is a computationally expensive function and hence the magnitude is itself often approximated by the more efficient and simpler technique of Equation (6.19):

$$|\nabla I| = |I \otimes G_x| + |I \otimes G_y|. \tag{6.19}$$

Equation (6.19) yields a surprisingly concise and efficient method for computing the MoG. In this equation, the MoG is given in terms of convolution and image addition. Listing 6.13 shows a complete implementation for computing the MoG of a source image. In this implementation, the source image is first convolved with the horizontal and vertical gradient kernels, as shown in lines 3–10, taking the absolute value in the process. Finally, the two gradients are added and the result returned.

A very simple approach to edge detection is to compute the MoG and choose a threshold value such that all magnitudes greater than the threshold are designated as an edge boundary. Figure 6.17 gives an example where the magnitude of the gradient is shown as an image. This image is thresholded on the intensity channel to produce a binary image where white pixels indicate the presence of an edge and black pixels indicate no edge. While thresholding is easily accomplished, choosing the correct threshold is problematic. If the threshold is too high, significant edges will not be detected and if the threshold is set too low, the technique will be susceptible to image noise and spurious edges will be identified.

Edge detection is an active field of research in computer vision and image processing and many different techniques have been developed. Most of these techniques are either first or second-derivative-based. The first derivative techniques generally compute the MoG and include some consideration of the orientation as well. The second derivative-based techniques search for zero-crossings of the second derivative of an image. Gaussian noise reduction is usually employed as a preprocessing step in order to improve the noise immunity. Edge detection techniques mainly differ in the types of smoothing filters that are applied and the way the edge strengths are computed. Since many edge detection methods rely on the image gradients, they also differ in the types of filters used for computing gradient estimates. The Canny technique is perhaps the most effective edge-detection filter and is a first-order derivative technique that factors in the orientation of an image as well as the magnitude of the gradient.

The magnitude of the gradient can be inverted to construct an edge map. An edge map simply shows edges in an image as dark lines on a generally light background. Figure 6.18 gives an example of various edge maps for the Roberts and Prewitt operators.

The magnitude of gradient can be used to create interesting artistic filters. To create image outlines, for example, an edge map can be generated from a source image and then combined with the original source by using a minimum binary op or even by adding. A glow filter results from combining a source image

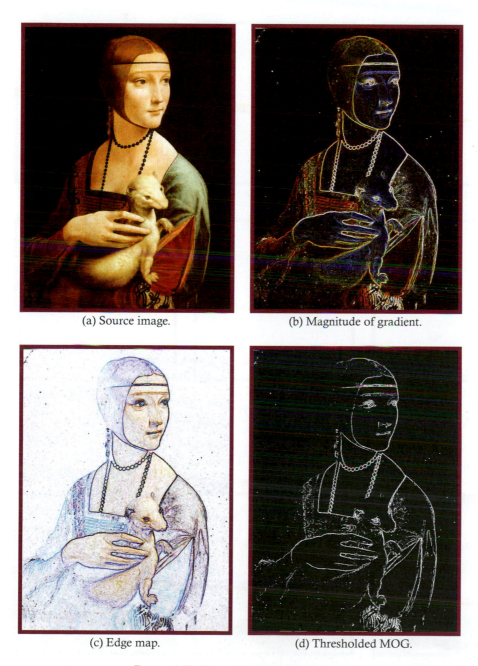

(a) Source image.   (b) Magnitude of gradient.

(c) Edge map.   (d) Thresholded MOG.

Figure 6.17. Edge detection with MoG.

(a) Roberts MoG.   (c) Roberts G_1.   (e) Roberts G_2.

(b) Prewitt MoG.   (d) Prewitt G_x.   (f) Prewitt G_y.

Figure 6.18. Edge maps using Prewitt and Roberts kernels.

(a) Source image.   (b) Texture.   (b) Glow.

Figure 6.19. Artistic effects using edges.

with its MoG via the maximum binary op. In either case, the edge map or the MoG can be blurred prior to recombining with the source such that the degree of blurring controls the strength of the corresponding artistic effect. Figure 6.19 gives an example of this technique.

6.5   Edge Enhancement

Edge enhancement, often referred to as sharpening, is a way of boosting the high frequency components of an image rather than, as edge detection techniques do, suppressing the low frequency components. Edge enhancement is often used to post-process images that have been acquired by scanning a document or painting.

Just as edge detection can be expressed as convolution, edge enhancement can also be achieved through convolution using an appropriately formed kernel. The following discussion shows how a suitable kernel can be obtained for sharpening. An image I is comprised of low frequency components, I_{low} and high frequency components I_{high}:

$$I = I_{low} + I_{high}. \tag{6.20}$$

Since edge enhancement seeks to increase the high frequency components, an edge enhanced image I_{sharp} can be formed by adding a portion of the high frequency component I_{high} to the source as in

$$I_{sharp} = I + \alpha I_{high}, \tag{6.21}$$

where the parameter α is a non-negative scaling factor controlling the overall degree of sharpening. Equation (6.20) can be solved for I_{high} and substituted into Equation (6.21) to obtain

$$\begin{aligned} I_{sharp} &= I + \alpha I_{high} \\ &= I + \alpha(I - I_{low}) \\ &= (1 + \alpha)I - \alpha I_{low}. \end{aligned} \tag{6.22}$$

The definition of sharpening given in Equation (6.22) will allow us to construct an appropriate convolution kernel. Consider a 3×3 convolution kernel $K_{identity}$ such that $I = I \otimes K_{identity}$ and a 3×3 kernel K_{low} such that $I_{low} = I \otimes K_{low}$. Equation (6.22) can then be expressed in terms of convolution with a sharpening kernel.

$$\begin{aligned} I_{sharp} &= (1 + \alpha)I - \alpha I_{low} \\ &= (1 + \alpha)(I \otimes K_{identity}) - \alpha(I \otimes K_{low}) \\ &= I \otimes (1 + \alpha)K_{identity} - I \otimes (\alpha K_{low}) \\ &= I \otimes ((1 + \alpha)K_{identity} - \alpha K_{low}). \end{aligned} \tag{6.23}$$

Equation (6.23) defines a sharpening kernel composed by subtracting a properly scaled blurring kernel from a scaled identity kernel. For the 3×3 case, the identity kernel contains zero-valued coefficients except for the center, which is unity. As previously described, a uniform low-pass kernel contains uniform positive coefficients. These kernels are shown below in normalized form:

$$K_{\text{identity}} = \frac{1}{9} \times \begin{bmatrix} 0 & 0 & 0 \\ 0 & 9 & 0 \\ 0 & 0 & 0 \end{bmatrix},$$

$$K_{\text{low}} = \frac{1}{9} \times \begin{bmatrix} 1 & 1 & 1 \\ 1 & 1 & 1 \\ 1 & 1 & 1 \end{bmatrix},$$

$$(1 + \alpha)K_{\text{identity}} - \alpha K_{\text{low}} = \frac{1}{9} \times \begin{bmatrix} -\alpha & -\alpha & -\alpha \\ -\alpha & (9 + 8\alpha) & -\alpha \\ -\alpha & -\alpha & -\alpha \end{bmatrix}.$$

The parameter α can be considered a gain setting that controls how much of the source image is passed through to the sharpened image. When $\alpha = 0$ no additional high frequency content is passed through to the output such that the result is the identity kernel. Figure 6.20 gives an example of edge enhancement showing two different gain values. This image is from a digitally scanned rendition of an oil painting by the renowned artist Eugene Delacroix. The image edges are enhanced through application of the sharpening filter as are the defects in the canvas. Higher scaling factors correspond to stronger edges in the output.

(a) Source image.   (b) $\alpha = .5$.   (c) $\alpha = 2.0$.

Figure 6.20. Image sharpening.

6.6 Rank Filters

Rank filtering is a nonlinear process that is based on a statistical analysis of neighborhood samples. Rank filtering is typically used, like image blurring, as a preprocessing step to reduce image noise in a multistage processing pipeline. The central idea in rank filtering is to make a list of all samples within the neighborhood and sort them in ascending order. The term *rank* means just this, an ordering of sample values. The rank filter will then output the sample having the desired rank.

The most common rank filter is the median filter where the output is the item having the median rank; the middle item in the sorted list of samples. For an $M \times N$ rectangular neighborhood there are MN samples in the neighborhood. When sorted, these samples have indices $[0, MN - 1]$ and the median sample has an index, or rank, of $\lfloor MN/2 \rfloor$ if we assume that that both M and N are odd.

Although median filtering and averaging via a uniform box convolution appear similar, they are actually very different. When computing the average value of a data set, every sample affects the resulting average. Samples that are extremely different than the median, and hence likely to be either an edge or noise, affect the result. When computing the median of a data set, although there is a sense in which all samples affect the output, it is also true that rogue samples have no real effect on the median. Median filtering is hence superior to averaging with respect to the task of eliminating noisy pixels while preserving edges or other fine image details.

Salt-and-pepper noise occurs when samples of an image are incorrectly set to either their maximum (salt) or minimum (pepper) values. Median filters are exceptionally good at reducing this type of noise since they do not smear the noise across the image as blurring filters do. Salt and pepper noise is common in imaging systems that use CMOS or CCD based image sensors. Since an image sensor is composed of millions of photo-sensors packaged within a small form factor, nearly every image sensor has a few defective photosites that continuously output either black or white regardless of the scene being imaged. Imaging systems often correct for such hardware defects by applying a median filter to the raw image data prior to shipping the image to the end user.

Figure 6.21 shows a numeric example of median filtering and illustrates how noisy samples are affected. In this example a 3×3 rectangular neighborhood is used to filter the sample in the center of the region that is most likely noise since it deviates significantly from the surrounding values. The nine samples falling within this region when sorted in ascending order are given as $\{50, 52, 53, 55, 59, 60, 61, 65, 253\}$. While the average of these samples is 78.6, the element having median rank is 59 which serves as the output as shown by the center pixel of part (b). In this instance, the value 59 is far more representative of the 3×3 region than the average value of 78.6 and hence is a better replacement for the noisy value than the average. When the entire source image

51	55	59	59	69
48	53	60	61	72
43	52	253	65	70
39	50	55	59	68
35	49	58	48	57

•	•	•	•	•
•	53	59	65	•
•	52	59	65	•
•	50	55	59	•
•	•	•	•	•

(a) Source image.   (b) Median filtered.

Figure 6.21. Numeric example of median filtering.

is median filtered, the process is repeated for every source sample to obtain the image of part (b).

Figure 6.22 gives an example of using median filtering to correct for impulse noise in a source image. In this example 60% of an images samples were deliberately set to either black or white in a randomly determined fashion in order to simulate a massive degree of salt-and-pepper noise. After median filtering the noise is significantly reduced and the image appears visually clean as seen in part (b). If Gaussian blurring is used to eliminate or reduce the noise, the result is to simply smear the noise across larger subregions of the image and produce the effect shown in part (c).

Unlike most other regional processing techniques, rank filtering often uses non-rectangular regions where the most common of such regions are in the shape of either a plus (+) or a cross (x). A rectangular mask is used to specify non-rectangular regions where certain elements in the region are marked as

(a) Noisy source image.   (b) Median filtered.   (c) Gaussian smoothed.

Figure 6.22. Median filtering.

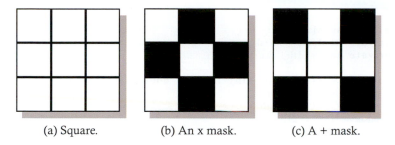

(a) Square.   (b) An x mask.   (c) A + mask.

Figure 6.23. Rank filtering masks for a 3×3 square, 3×3 x, and 3×3 +.

being included and others excluded from the region. Figure 6.23 gives examples of masks where the black entries signify that corresponding source samples are not included in the region while white entries denote source samples that are included in the region. If the black entries are understood as opaque blocks and the white entries as transparent blocks then the neighborhood defined by the mask includes all samples that visible through the transparent blocks of the mask.

Other rank-based filters include the minimum, maximum, and range filters. As indicated by their names, the minimum filter outputs the minimum value of all neighborhood samples, the maximum filter outputs the maximum value, and the range filter outputs the difference (the range) between the maximum and minimum values.

6.6.1 Implementation

A general-purpose rank filter can be written that also supports non-rectangular regions. Such a filter can serve as either a minimum, median, or maximum filter or a filter of any rank depending on the parameters provided when the filter is applied. It should be noted that since rank filtering is nonlinear it cannot be implemented using convolution.

In order to support arbitrarily shaped regions, a Mask class is developed. Listing 6.14 implements a mask where a linearized array of booleans in row-major layout is used to represent an $M \times N$ rectangular region of an image. In this listing, a true value indicates inclusion in the neighborhood and a false value indicates that the corresponding sample is not included in the neighborhood. The key element (the center element) is given by coordinate $(0, 0)$ where both M and N are generally required to be odd valued in order for the mask to have a properly centered key element. Elements of the mask are accessed via the isIncluded method, which computes whether an element is within the region or not. The mask is constructed in similar fashion to that of the Kernel class where the boolean values are given as a linear array and the dimensions of the region are given as integer-valued parameters.

```
1  public class Mask {
2    private boolean[] mask;
3    private int width, height;
4
5    public Mask(int width, int height, boolean[] flags) {
6      this.width = width;
7      this.height = height;
8      mask = flags.clone();
9    }
10
11   public int getWidth() { return width; }
12
13   public int getHeight() { return height; }
14
15   // element (0,0) is the center element
16   // and hence indices x and y may be negative
17   public boolean isIncluded(int x, int y) {
18     return mask[(y+height/2)*width+(x+width/2)];
19   }
20
21   public int getSample(int x, int y) {
22       return mask[(y+height/2)*width+(x+width/2)] ? 255 : 0;
23   }
24
25   public int getSize() {
26     int count = 0;
27     for(int i=0; i<mask.length; i++) {
28       if(mask[i]) count++;
29     }
30     return count;
31   }
32
33   public Rectangle getBounds() {
34     return new Rectangle(-width/2, -height/2, width, height);
35   }
36
37   public Mask invert() {
38       Mask result = new Mask(width, height, mask);
39       for(int i=0; i<mask.length; i++){
40           result.mask[i] = !mask[i];
41       }
42       return result;
43   }
44 }
```

Listing 6.14. Mask.

It is often necessary to know how many samples are actually included in the mask and hence the getSize method is provided as a way of computing the number of true values within the region. When using a Mask for rank

```
1  public class RankOp extends NullOp {
2    private Mask mask;
3    private int rank;
4    private ImagePadder padder = ExtendedBorderPadder.getInstance();
5
6    public RankOp(Mask m, int r) {
7      if (r >= m.getSize() || r < 0) {
8        throw new IllegalArgumentException("invalid␣rank:␣" + r);
9      }
10     rank = r;
11     mask = m;
12   }
13
14   public void setRank(int r) {
15     rank = r;
16   }
17
18   public void setMask(Mask m) {
19     mask = m;
20   }
21
22   private int getSampleOfRank(BufferedImage src, Location pt, int[] list) {
23     int i = 0;
24     for(Location mPt : new RasterScanner(mask.getBounds())) {
25       if(mask.isIncluded(mPt.col, mPt.row)) {
26         list[i++] = padder.getSample(src,
27                                      pt.col + mPt.col,
28                                      pt.row + mPt.row,
29                                      pt.band);
30       }
31     }
32
33     Arrays.sort(list);
34     return list[rank];
35   }
36
37   @Override
38   public BufferedImage filter(BufferedImage src, BufferedImage dest) {
39     if (dest == null) {
40       dest = createCompatibleDestImage(src, src.getColorModel());
41     }
42     int[] list = new int[mask.getSize()];
43     for (Location pt : new RasterScanner(src, true)) {
44       int output = getSampleOfRank(src, pt, list);
45       dest.getRaster().setSample(pt.col, pt.row, pt.band, output);
46     }
47     return dest;
48   }
49 }
```

Listing 6.15. General purpose rank filter.

filtering, the rank must then be in the range [0, getMaskSize()-1]. The invert method creates a new mask that is the logical inverse of the active object.

Listing 6.15 is a complete implementation of a general-purpose rank filter that uses a Mask to determine the region's shape. A RankOp is constructed by passing in a Mask along with a rank value. The class allows clients to change both the rank and the mask via the setRank method and setMask methods, respectively. The *filter* method corresponds to the maximum filter when rank=mask. getMaskSize()-1, to the median filter when rank=mask.getMaskSize()/2, and to the minimum filter when rank=0.

The filter method iterates over every sample of the source and applies rank filtering for each sample via the getSampleOfRank method. The getSample OfRank method applies the mask to a single sample within the source image where the sample is specified with a Location object. The method also accepts an array name *list*. This array does not contain any information but is simply a reusable data structure that is created once by the client rather than recreating an array with every call to getSampleOfRank. The array is created on line 42 and simply passed into the function on line 44. For each sample in the surrounding region, the mask is queried to determine whether or not the sample is included,

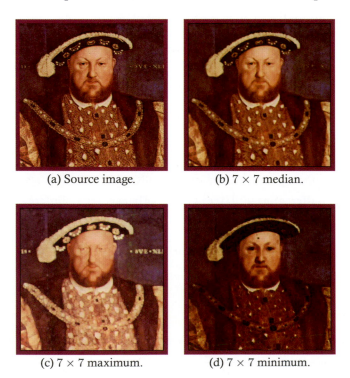

(a) Source image.   (b) 7 × 7 median.

(c) 7 × 7 maximum.   (d) 7 × 7 minimum.

Figure 6.24. Rank filtering examples.

as shown in line 25. If it is, then the sample is added to the list, which is finally sorted via the standard `Arrays` class and the proper element is returned.

Figure 6.24 gives examples of rank filtering a source image using various ranks. The source image of (a) is filtered with a 7×7 median filter in (b), a 7×7 maximum rank filter in (c), and a 7×7 minimum filter in (d).

6.6.2 Analysis

The general purpose rank filter of Listing 6.15 is a computationally slow process since it requires sorting a list *for every sample* of a source image. For ease of analysis assume that the neighborhood is an $N \times N$ rectangular region and that source image I is a $W \times H$ grayscale image. For each sample of the source image a list of size N^2 must be created and then sorted. If we assume that sorting itself, as given by the `Arrays.sort` method call of Listing 6.15, takes time proportional to $N^2 \log(N^2)$ the brute-force approach to rank filtering takes time proportional to $WH(N^2 \log(N^2))$. This implementation is impractically slow since it grows quadratically with respect to the size of the mask.

Significant savings can be obtained by writing specialized minimum and maximum filters rather than using the generalized rank filter. Finding the minimum or maximum value of a list does not require sorting; a linear scan through the region is sufficient. Modifying the operation accordingly will result in a runtime proportional to WHN^2.

Even median filtering can be made more efficient by approximating the median computation. If the region is a square $N \times N$ mask, an approximate median can be computed by by successive applications of two linear median filters. When finding the approximate median, a source image I is first median filtered using a $N \times 1$ mask to produce the median value of each row within the neighborhood. This intermediate result is then median filtered using a $1 \times N$ mask to produce an approximate median filtered result. This approximation is

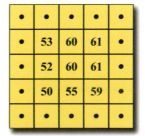

(a) Source image.   (b) Row approximation.   (c) Final approximation.

Figure 6.25. Numeric example of fast (approximate) median filtering.

illustrated in Figure 6.25, where a 5×5 source image is median filtered using a 3×3 mask. An intermediate image is constructed by median filtering with a 3×1 mask to create the intermediate result shown in (b). This intermediate image is then median filtered using a 1×3 mask to generate the final approximation.

The use of separate masks when performing approximate median filtering is directly analogous to the use of separable kernels when performing convolution and results in a similar increase in performance. Exact median filtering can be performed more efficiently, however, by a more sophisticated approach that uses radix sort in conjunction with the fact that adjacent regions overlap to a large extent and hence the computation of their respective medians can be shared between the regions [Huang et al. 79] [Ataman et al. 80].

The algorithm of Listing 6.16 show how fast median filtering can be done using a histogram to sort the neighborhood samples. A raster scan is performed and the first sample in each row essentially resets the process. The histogram of the first sample of each row is created and then used to find the median value. For a square $N \times N$ region, the median value is given by the smallest m such that $\sum_{i=0}^{m} H(i) >= \lceil N^2/2 \rceil$. In other words, the median is the first histogram index for which the sum of the values to the index reaches $\lceil N^2/2 \rceil$. Given the regional histogram, the median can be found by finding the running sum of histogram elements until the sum exceeds the specified threshold. While computation of the median value for the first column in each row is not efficient, each subsequent sample in the row can be determined with great efficiency by the careful updating of the histogram.

For every subsequent sample in the row the histogram is updated by subtracting one count for every sample leaving the region and adding a count for every

```
1  Image algorithm fastMedian(Image src, int N) {
2    DEST = destination image
3
4    for every row R of the src (top-to-bottom)
5      Set H to be the histogram of the NxN region of first sample
6      find the median value MED by examining H
7      write MED to DEST
8      for every remaining column C of R (left-to-right)
9        Update H by removing the value in the left most column
10             of the previous region
11       Update H by adding the values in the right most column
12             of the current region
13       find the median value MED by scanning higher or lower depending on
14         the values that have been removed/added
15
16       write MED to DEST
```

Listing 6.16. Fast median filtering.

sample entering the region. After the histogram has been so updated, the median value can be found by scanning either higher or lower in the new histogram depending on whether the alteration to the histogram forced the median lower or higher than the previous value. This process can be further improved by using a serpentine scan such that resetting the histogram at the beginning of each row is not necessary.

6.6.3   Other Statistical Filters

Range filtering is related to rank filtering but it is not itself a rank filter and hence cannot be accomplished directly by the RankOp class. Range filtering produces the *difference* between the maximum and minimum values of all samples in the neighborhood such that it generates large signals where there are large differences (edges) in the source image and will generate smaller signals where there is little variation in the source image.

Range filtering can be implemented by a slight modification to the code of Listing 6.15 or by using the RankOp in conjunction with image subtraction. Let image $I'_{\max}$ be the result of applying a maximum filter to source image I and let $I_{\min}$ be the result of applying a minimum filter to the same source

(a) Most common sample.   (b) Range filtering.

Figure 6.26. Other statistically-based filters.

image. The range filtered image can then be computed using image subtraction as $I_{\text{range}} = I_{\text{max}} - I_{\text{min}}$. Range filtering is a type of relatively fast but generally poor performing edge detection filter.

Rather than keying on rank it is possible to select the most common element in the neighborhood. This is a histogram-based approach that produces a distorted image where clusters of color are generated. The resulting image will generally have the appearance of a mosaic or an oil painting with the size and shape of the mask determining the precise effect. If two or more elements are most common the one nearest the average of the region should be selected; otherwise choose the darker value. It should be noted that performing this filter in one of the perceptual color spaces produces superior results since smaller color distortions are introduced. Figure 6.26 gives examples of both range and histogram-based filtering where the image of (a) has been processed using the most common sample technique with 17×17 square region and then sharpened, and the image of (b) has been range filtered with a 5×5 square mask.

6.7   Template Matching and Correlation

Consider the problem of locating an object in some source image. In the simplest case the object itself can be represented as an image which is known as a template. A template matching algorithm can then be used to determine where, if at all, the template image occurs in the source.

One template matching technique is to pass the template image over the source image and compute the error between the template and the subimage of the source that is coincident with the template. A match occurs at any location where the error falls below a threshold value. Given a source image f and an $M \times N$ template K where both M and N are odd we can compute the root mean squared (RMS) error at every source location (x, y). Since the RMS error is computed for every source sample, we obtain an output image where dark pixels (low error values between the template and the source) signify a strong likelihood of a match and bright pixels (large error values) signify that there is no match. The RMS error can be computed as given in Equation (6.24), where g is the resulting destination image.

$$g(x,y) = \sqrt{\frac{1}{MN} \sum_{j=-\lfloor \frac{M}{2} \rfloor}^{\lfloor \frac{M}{2} \rfloor} \sum_{k=-\lfloor \frac{N}{2} \rfloor}^{\lfloor \frac{N}{2} \rfloor} [f(x+j, y+k) - K(j,k)]^2}, \qquad (6.24)$$

A slightly more sophisticated technique is known as cross correlation which is given in Equation (6.25). Observant readers will note the similarity between this formula and that of convolution given in Equation (6.1), where the only difference is given by the sign preceding offsets j and k. You may recall that under convolution the kernel (or template) is rotated while Equation (6.1) indicates that under correlation the kernel is not rotated. The template is generally not designated as a *kernel* since a template is generally much larger than a kernel and is of the same nature as the source, having coefficient values of the same range and color depth.

$$g(x, y) = \sum_{j=-\lfloor M/2 \rfloor}^{\lfloor M/2 \rfloor} \sum_{k=-\lfloor N/2 \rfloor}^{\lfloor N/2 \rfloor} K(j, k) \cdot I(x + j, y + k). \qquad (6.25)$$

Correlation as given in Equation (6.25) is problematic for use in template matching since it does not take into account the relative intensity of either the source or the template. Consider searching for an all-black region within an all-black source image. The resulting cross correlated destination will itself be all-black, presumably indicating exact matches at every source location. Now consider searching for an all-white region within an all-white source image. The resulting cross correlated destination will be all-white, presumably indicating that no matches exist. This gives rise to the question of how to identify matches within the output image if a match could be indicated by either dark or bright regions.

A solution can be obtained by normalizing the output according to the relative brightness of both the template and the source images. This technique is known as normalized cross correlation and is given in Equation (6.26), where offsets j and k range over the entire width and height of the template. $\overline{K}$ is the average value of all samples in the template and can be precomputed once while $\overline{I}(x, y)$ is the average value of all samples spanning the template at location (x, y) of the source and must be computed for every location within the source:

$$g(x, y) = \frac{\sum_j \sum_k [K(j, k) - \overline{K}] \cdot [I(x + j, y + k) - \overline{I}(x, y)]}{\sqrt{\sum_j \sum_k [K(j, k) - \overline{K}]^2 \sum_j \sum_k [I(x + j, y + k) - \overline{I}(x, y)]^2}}. \qquad (6.26)$$

Figure 6.27 illustrates how template matching can be used to locate objects within a source image. In this example we are searching the source image for all occurrences of a red diamond shape lying on a black background. The template image itself, shown as an inset within the source image, was taken from the right thigh of the harlequin figure and is thus an exact match in one location of the image. The red diamond pattern, however, is also found to be very similar to other regions of the image as indicated by the dark regions of the RMS technique. The RMS errors are shown in the grayscale image of part (b) as the average of the errors over the three color bands. Thresholding part (b) to identify the lowest RMS regions results in the likely matches shown in part (c).

Template matching is used in many applications to locate image features. Biologists may use template matching to identify moving cells on a microscopic

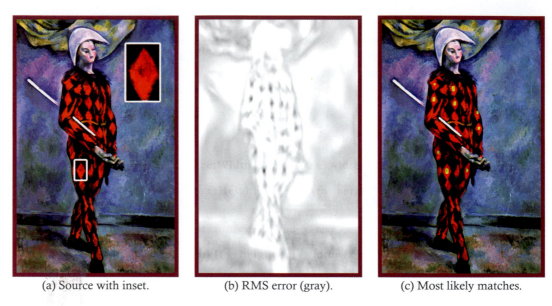

(a) Source with inset.   (b) RMS error (gray).   (c) Most likely matches.

Figure 6.27. Template matching.

image, meteorologists may track the movement of hurricanes, and optical character recognition may utilize correlation to identify individual characters on a binary image of printed text.

Correlation is sensitive to both the scale and rotation of the template. If the template object is not exactly the same size and orientation as similar objects in the source, then template matching will not be effective and alternate techniques should be used. In addition, the computational cost of correlation is high since the template is generally large. Frequency based implementations should generally be used and are discussed Chapter 9.

6.8   Exercises

1. Convolve a set of test images with the following convolution kernels. You should first anticipate the visual effect that the kernels will produce and compare your expectations with the actual result:

$$\frac{1}{6} \times \begin{bmatrix} 0 & -1 & 0 \\ -1 & 10 & -1 \\ 0 & -1 & 0 \end{bmatrix} \quad \frac{1}{3} \times \begin{bmatrix} -1 & 0 & -1 \\ 0 & 7 & 0 \\ -1 & 0 & -1 \end{bmatrix}.$$

2. Write a class named `ImpulseNoiseOp` that implements the `Buffered ImageOp`. This class should filter an image by adding impulse noise to

the source. The `ImpulseNoiseOp` is useful for testing noise reduction techniques. In addition to the required methods of the `BufferedImageOp` the class should support those methods shown in the UML class diagram below. The amount is a value in the range $[0, 1]$ and is meant to indicate the percentage of samples that will be corrupted by impulse noise. When filtering a source image this class should generate a random value R in $[0, 1]$ for every source sample. If $R <=$ amount then the corresponding destination sample should be randomly set to either black or white. If $R >$ amount then the source sample is copied to the destination.

≪ *implements   BufferedImageOp* ≫
ImpulseNoiseOp
+ImpulseNoiseOp(amount:double)
+void setAmount(amount:double)
+void setAmount(amount:double)
+double getAmount():double

3. Write specialized `MinimumOp` and `MaximumOp` filters that perform efficient minimum and maximum rank filtering as described in Section 6.6.2. Compare the performance of your classes by filtering a large source image with your specialized filters and the general rank filter of Listing 6.15. Compare the time taken by each implementation.

4. Write an `ApproximateMedianOp` filter that performs approximate median filtering. Compare the efficiency of this implementation with the general `RankOp` filter on a number of images having large dimension and plot the results. Your plot should show the length of time taken on images of various sizes using your filter and the `RankOp` filter of the text.

5. Write a `RangeOp` filter as described in Section 6.6.3.

⋆ 6. Write a `FastMedianOp` filter that performs median filtering using the histogram-based method presented in Listing 6.16.

⋆ 7. Write a `MostCommonRegionalOp` that computes the most common sample in a region as the output of a regional processing operation as described in Section 6.6.3. Your filter should work in the HSB color space.

⋆⋆ 8. Write an `NCCOp` that computes the normalized cross correlation of an image with a template image.

Artwork

Figure 6.5. *Young Italian Woman at a Table* by Paul Cézanne (1839–1906). Paul Cézanne was a French Post-Impressionist painter. His work formed a bridge between late 19th century Impressionism and the Cubism of the early 20th century. Cézanne was a master of design, composition, and color using bold strokes of color to highlight personality and presence. The portrait detail of Figure 6.5 was completed around 1885–1900 during a period that saw him produce several such masterful figure studies.

Figure 6.10. *Study for the Head of Leda* by Leonardo da Vinci (1452–1519). Leonardo da Vinci has often been described as the most talented *Renaissance man* of all ages. He is widely considered to be one of the greatest painters of all time in addition to his numerous accomplishments as mathematician, physicist, and engineer. The drawing shown in Figure 6.10 is the only existing evidence of a painting that was likely completed and subsequently destroyed or lost. The subject of Leda and the swan is drawn from Greek mythology and was popular in da Vinci's day.

Figure 6.12. *Still Life with Pomegranates and Pears* by Paul Cézanne (1839–1906). See the entry of Figure 6.5 for a short biographical sketch of the artist. The still life of Figure 6.12 is representative of his style.

Figure 6.15. *Portrait of a Woman Holding a Fan* by Katsushika Hokusai (1760–1849). Katsushika Hokusai was a Japanese artist, ukiyo-e painter, and printmaker. He is best known for his woodblock prints and is perhaps the most famous for *The Great Wave off Kanagawa*, which was produced in the 1820s. Hokusai enjoyed a long career but produced his best work after the age of 60. In fact, he is reported to have said near the time of his death that "If only Heaven will give me just another ten years... Just another five more years, then I could become a real painter."

Figure 6.16. *Portrait of a Woman Holding a Fan* by Katsushika Hokusai (1760–1849). See the entry of Figure 6.15.

Figure 6.17. *Lady with an Ermine* by Leonardo da Vinci (1452–1519). See the entry of Figure 6.8 for a short biographic sketch of the artist. The portrait painting of Figure 6.17 was completed around 1489. The lady of the portrait is most likely Cecilia Gallerani, the mistress of Lodovico Sforza, the Duke of Milan. The painting is one of only four female portraits Leonardo painted and despite sustaining significant damage over time is nonetheless one of the most well preserved of da Vinci's paintings. At the time of painting, Ms. Gallerani was about 17 years old and was a gifted musician and poet. It appears as if the ermine is actually a white ferret, the significance of which is uncertain.

Figure 6.18. *Portrait of a Woman Holding a Fan* by Katsushika Hokusai (1760–1849). See the entry of Figure 6.15.

Figure 6.19. *Peasant Woman from Muscel* by Nicolae Grigorescu (1838–1907). Grigorescu was born in Romania where, as a young man, he became an apprentice of the painter Anton Chladek creating icons for the church of Baicoi and the monastery of Caldarusani. He later moved to Paris and became acquainted with colleagues such as Pierre-August Renoir, Jean-François Millet and Gustave Courbet among others. His later works emphasized pastoral themes and peasant women of which Figure 6.19 is representative.

Figure 6.20. *Orphan Girl at the Cemetery* by Ferdinand Victor Eugéne Delacroix (1798–1863). Delacroix was a French Romantic painter who from early in his career was viewed as one of the best leaders of the French Romantic school. Delacroix was also an excellent lithographer and illustrated various literary works including those of Shakespeare, Goethe, and others. Delacroix was heavily influenced by the art of Rubens and other Renaissance painters who emphasized color and movement rather than realism and clarity of form. Figure 6.20 gives a detail of a portrait that he completed in 1824.

Figure 6.22. *Portrait of Henry VIII* by Hans Holbein the Younger (1497–1543). Hans Holbein was a German artist of the Northern Renaissance style. He was raised by his father, Hans Holbein the Elder, who was himself an accomplished artist and trained his son in the arts from an early age. Holbein traveled to London in 1526 and was eventually commissioned to complete numerous portraits at the court of Henry VIII. Among his subjects were Anne Boleyn, Jane Seymour, Anne Cleves, King Edward VI and, as is shown in the detail of Figure 6.24, Henry VIII. Holbein apparently died of the plague while working to finish another portrait of Henry VIII.

Figure 6.24. *Portrait of Henry VIII* by Hans Holbein the Younger (1497–1543). See the entry of Figure 6.20.

Figure 6.26. *Orphan Girl at the Cemetery* by Ferdinand Victor Eugéne Delacroix (1798–1863). See the entry of Figure 6.22.

Figure 6.27. *Harlequin* by Paul Cézanne (1839–1906). See the entry of Figure 6.5 for a short biographical sketch of the artist.

Geometric Operations

7

Most of the operations that we have so far discussed process pixels by simply changing their color. In this chapter we will discuss operations that leave the color of a pixel intact but change the *location* of pixels. These operations are known as geometric operations since they affect change on the spatial structure, or geometry, of an image.

Image registration is a geometric operation that takes multiple images of a single scene that are shot at different angles and combines them into a single coherent whole. The source images may have been acquired by completely different imaging systems or by a single system taking a series of images over time. It is common in medical imaging, for example, to combine a series of images from sources such as x-ray, MRI, and NMR scans into a single visual image. The source images are carefully aligned, or registered, by correlating the geometry of the subject among the various sources. The resulting composite provides greater resolution and more thorough detail than any of the source images provides in isolation.

Image registration has also been used to produce panoramic composites of the surface of Mars where the individual images were acquired by the same system planetary explorer over a series of time by simply rotating the camera across a certain field of view.

Geometric operations are also able to correct hardware defects or to otherwise compensate for limitations in the image acquisition system. The Hubble space telescope, for example, was initially deployed with such a flawed optical system that scientists feared the satellite would not be able to provide usable data of any real significance. The flaw was fixed through a combination of replacing the defective optics and defect-correcting image processing software fine-tuned for the Hubble system's defects.

Geometric operations generally operate by shifting pixel locations from their source positions to more appropriate positions in the destination. Image resampling, rotating, shearing, reflection, and morphing are examples of operations that directly affect the geometry of an image.

7.1 Affine Transformations

In general, geometric operations take a source pixel at some location (x, y) and map it to location (x', y') in the destination. The mapping between (x, y) and (x', y') can be generalized as defined in Equation (7.1), where both T_x and T_y are transformation functions that produce output coordinates based on the x and y coordinates of the input pixel. Both functions produce real values as opposed to integer coordinates and are assumed to be well defined at all locations in the image plane. Function T_x outputs the horizontal coordinate of the output pixel while function T_y outputs the vertical coordinate:

$$
\begin{aligned}
x' &= T_x(x, y), \\
y' &= T_y(x, y).
\end{aligned}
\tag{7.1}
$$

The simplest category of transformation functions is linear in (x, y) and is known as affine transformations. In other words, whenever the transformation functions are linear functions of variables x and y the transform is affine. Equation (7.2) defines the entire family of affine transformations where the m values are scalar real-valued coefficients. The central properties of an affine transformation are that straight lines within the source remain straight lines in the destination and that parallel lines within the source remain parallel in the destination.

$$
\begin{aligned}
T_x(x, y) &= m_{00}x + m_{01}y + m_{02}, \\
T_y(x, y) &= m_{10}x + m_{11}y + m_{12}.
\end{aligned}
\tag{7.2}
$$

These two equations are often augmented by a third equation that may initially seem frivolous but allows for convenient and efficient computation. The validity of the equation $1 = 0x + 0y + 1$ is obvious but including it as a third constraint within our system allows us to write the affine system in the matrix form of Equation (7.3). The 3×3 matrix of coefficients in Equation (7.3) is known as the *homogeneous transformation matrix*. Using this augmented system, a two-dimensional point is represented as a 3×1 column vector where the first two elements correspond to the column and row while the third coordinate is constant at 1. When a two-dimensional point is represented in this form it is referred to as a homogenous coordinate. When using homogenous coordinates, the transformation of a point V with a transformation matrix A is given by the product AV. In other words, multiplication of a transformation matrix with a point yields a point:

$$
\begin{bmatrix} x' \\ y' \\ 1 \end{bmatrix} = \begin{bmatrix} m_{00} & m_{01} & m_{02} \\ m_{10} & m_{11} & m_{12} \\ 0 & 0 & 1 \end{bmatrix} \cdot \begin{bmatrix} x \\ y \\ 1 \end{bmatrix}.
\tag{7.3}
$$

An affine transformation matrix is a six parameter entity controlling the coordinate mapping between source and destination images. Affine transformations are capable of performing image translation, rotation, scaling, shearing,

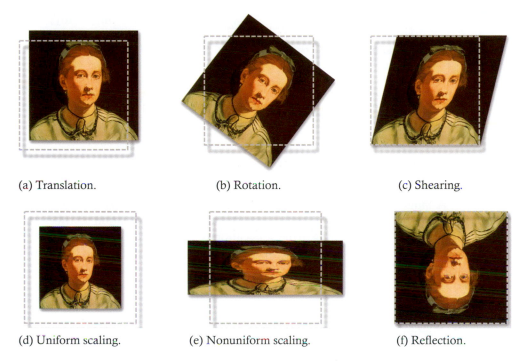

(a) Translation.   (b) Rotation.   (c) Shearing.

(d) Uniform scaling.   (e) Nonuniform scaling.   (f) Reflection.

Figure 7.1. Linear geometric transformations.

and reflection by appropriate selection of the six parameters. These geometric operations are illustrated in Figure 7.1. In each example the destination image is displayed relative to the bounds of the source image as shown by the dashed boxes.

Scaling adjusts the size of an image and can be uniform, using the same scaling factor in both the horizontal and vertical dimensions, or nonuniform where different scaling factors are used in the horizontal and vertical dimensions. Image translation simply repositions the image, or subimage, at a different location. An image can be rotated around any point; rotation about the image center is shown in this figure. Shearing can occur with respect to either the horizontal or vertical axis or both; horizontal shearing is illustrated here. Reflection negates the coordinates of pixels in an image around any user-specified axis but is typically performed around either the central horizontal or vertical axis.

Table 7.1 defines how each of the six coefficients of the homogeneous transformation matrix must be chosen to perform these affine operations. Rotation is parameterized by an angular amount θ; scaling is parameterized on the scaling factors s_x and s_y, while reflection is parameterized on the location of either the horizontal or vertical axes as defined by x_c or y_c.

	m_{00}	m_{10}	m_{01}	m_{11}	m_{02}	m_{12}
translation	1	0	0	1	δx	δy
rotation	$\cos(\theta)$	$-\sin(\theta)$	$\sin(\theta)$	$\cos(\theta)$	0	0
shear	1	sh_y	sh_x	1	0	0
scale	s_x	0	0	s_y	0	0
horizontal reflection	-1	0	0	1	x_c	0
vertical reflection	1	0	0	-1	0	y_c

Table 7.1. Coefficient settings for affine operations.

A single homogeneous matrix can also represent a *sequence* of individual affine operations. Letting A and B represent affine transformation matrices, the affine matrix corresponding to the application of A followed by B is given as BA, which is itself a homogeneous transformation matrix. Matrix multiplication, also termed concatenation, therefore corresponds to the sequential composition of individual affine transformations. Note that the order of multiplication is both important and opposite to the way the operations are mentally envisioned. While we speak of *transform A followed by transform B*, these operations are actually composed as *matrix B multiplied by (or concatenated with) matrix A*. Assume, for example, that matrix A represents a rotation of $-30°$ about the origin and matrix B represents a horizontal shear by a factor of .5; the affine matrix corresponding to the rotation followed by shear is given as BA. This is illustrated in Equation (7.4), where the leftmost matrix represents the shear matrix B which is preconcatentated to A to obtain the rightmost matrix BA. The matrix BA is a single affine transform that embodies in itself a rotation followed by a shear.

$$
\begin{bmatrix} 1 & 0.5 & 0 \\ 0 & 1 & 0 \\ 0 & 0 & 1 \end{bmatrix} \cdot \begin{bmatrix} .866 & -0.5 & 0 \\ 0.5 & .866 & 0 \\ 0 & 0 & 1 \end{bmatrix} = \begin{bmatrix} 1.116 & -0.067 & 0 \\ 0.5 & .866 & 0 \\ 0 & 0 & 1 \end{bmatrix}. \quad (7.4)
$$

Consider, for example, the question of how to rotate an image by $30°$ clockwise (or by $-\pi/6$ radians) where we understand that rotation using the transformation matrix is always done about the origin of the image. Table 7.1 indicates that we must form the affine transformation matrix, as shown in Equation (7.5) since $\cos(-\pi/6) = .866$ and $\sin(-\pi/6) = -1/2$:

$$
\begin{bmatrix} x' \\ y' \\ 1 \end{bmatrix} = \begin{bmatrix} .866 & -0.5 & 0 \\ 0.5 & .866 & 0 \\ 0 & 0 & 1 \end{bmatrix} \cdot \begin{bmatrix} x \\ y \\ 1 \end{bmatrix}. \quad (7.5)
$$

Under rotation, the pixel at location $(10, 20)$ must be relocated to another location in the destination. When the pixel at location $(10, 20)$ is transformed under the matrix of Equation (7.5) the pixel is relocated to position (x', y') in the destination which is given by Equation (7.6).

$$\begin{bmatrix} x' \\ y' \\ 1 \end{bmatrix} = \begin{bmatrix} .866 & -0.5 & 0 \\ 0.5 & .866 & 0 \\ 0 & 0 & 1 \end{bmatrix} = \begin{bmatrix} -1.34 \\ 22.32 \\ 1 \end{bmatrix}. \tag{7.6}$$

This example presents two important issues that arise when performing geometric transformations. The first issue relates to the dimensionality of the destination with respect to the source. The source sample at location $(10, 20)$, for example, ended up at location $(-1.34, 22.32)$ in the destination. Negative indices are not allowed in Java and hence care must be taken to shift the destination coordinates back into valid indices in the destination. The second issue relates to the details of how mapping is done since the destination coordinates are not integers but real values. Since real-valued locations are not allowed in Java, care must also be taken to appropriately interpret the location mappings. These two issues are discussed in the following subsections.

7.1.1 Destination Dimensionality

Figure 7.2 illustrates the sizing and positioning issues that arise when rotating an image. While rotation is used to motivate these issues, these are issues that must be addressed for all affine transformations. The source image is rotated about the origin such that some pixels are mapped outside of the bounds of the source image as illustrated by the middle image. Implementations must decide whether to size the destination image in such a way as to truncate the result or allow the destination to contain the entire rotated image.

In order to view the entire rotated image both the width and height of the destination image must be increased beyond that of the source. In addition, each

(a) Source image.

(b) Rotation.

(c) Expanded view of rotated image.

Figure 7.2. Destination dimensionality under rotation.

of the pixels must be translated onto the resized destination so that all coordinates are non-negative and integer valued. The rightmost image shows that the size of the destination is larger than the source and that pixels have been shifted. Note, for example, that source pixel $(0,0)$ is mapped to some coordinate $(x',0)$ where x' is a positive value.

The dimensions of the destination image can be determined by taking the four points defining the bounds of the source image, transforming them via the transformation matrix, and then finding the bounds of the result. Destination pixels can then be translated by the location of the upper-left coordinate of the bounding rectangle of the destination.

7.1.2   Mapping

If we assume that the destination image is adequately dimensioned, the issue of how integer-valued source coordinates are mapped onto integer-valued destination coordinates must also be addressed. One technique is known as forward mapping. Forward mapping takes each pixel of the source image and copies it to a location in the destination by rounding the destination coordinates so that they are integer values.

Forward mapping yields generally poor results since certain pixels of the destination image may remain unfilled as illustrated in Figure 7.3. In this example, a source image is rotated by $45°$ using a forward mapping strategy where the destination is sufficiently sized and the rotated image appropriately translated onto the destination.

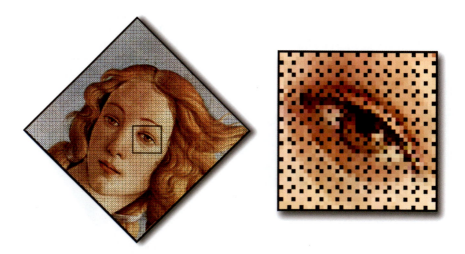

Figure 7.3. Rotation by forward mapping.

Perhaps the simplest way of understanding the problem introduced by forward mapping is to consider the scaling transformation. When increasing the scale or size of an image the destination image is larger than the source. If every source sample is copied over to the destination it is apparent that not all pixels of the destination will be filled. For this reason, forward mapping is rarely used for geometric transformations.

Backward mapping solves the gap problem caused by forward mapping. When using backward mapping an empty destination image is created and each location in the destination is mapped *backwards* onto the source. After identifying the source location that corresponds to the destination a sample value is obtained via interpolation.

Backward mapping is generally straightforward when using affine transformations. Assume that the 3×1 homogeneous coordinate $(x, y, 1)$ is transformed by a homogeneous transformation matrix A to obtain $(x', y', 1)$. Letting $v = [x, y, 1]^T$ and $v' = [x', y', 1]^T$ we can write $v' = Av$. Since matrix A and vector v are known when doing forward mapping, this formulation allows quick computation of v'. When backward mapping we scan over the destination locations and therefore wish to compute vector v given A and v'. In order to solve for v', the affine transformation matrix must be inverted and the equation must be solved for v. The solution is given in Equation (7.7), where the inverse of matrix A is denoted as A^{-1}:

$$
\begin{aligned}
v' &= Av, \\
A^{-1}v' &= A^{-1}Av, \\
A^{-1}v' &= v.
\end{aligned}
\tag{7.7}
$$

When performing backward mapping then, Equation (7.7) implies that we construct an affine transformation matrix that corresponds to the desired mapping and that this matrix is then inverted. Every pixel of the destination is scanned and the corresponding source location is computed by multiplying the inverted matrix by the destination location. This source sample is then copied from the source to the destination. Although most affine transformations are invertible, there are affine transformations[1] that are not, and hence, while backward mapping can usually be employed, it cannot be used for all possible affine transformations.

7.1.3 Interpolation

Even when using backward mapping, however, a destination pixel may map to a non-integer location in the source. Assume, for example, that we wish to determine the value of the destination sample $I'(x', y')$ using backward mapping.

---

[1] An affine transformation is invertible if it has a nonzero determinant.

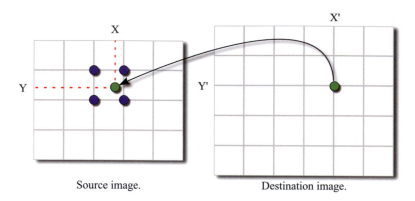

Figure 7.4. Reverse mapping.

Location (x', y') is transformed by the inverse affine transformation matrix to obtain the corresponding source location (x, y). Since the source location (x, y) will not generally lie on integer coordinates the problem is to determine the intensity value at that location. Figure 7.4 gives a pictorial illustration of this problem. In this figure, the image samples lie at the grid intersections and since the destination sample originates from between samples in the source, it is unclear as to which, if any, of the surroundings samples should be used to determine the value of the destination sample.

This problem is solved by using interpolation. Interpolation refers to the creation of new samples from existing image samples. In other words, interpolation seeks to increase the resolution of an image by adding virtual samples at all points within the image boundary. Recall that a sample represents the intensity of light at a single point in space and that the sample is displayed in such a way as to spread the point out across some spatial area. Using interpolation it is possible to define a new sample at any point in space and hence the use of real-valued coordinates poses no difficulty since we can infer an image sample at all locations.

The simplest interpolation technique is nearest neighbor interpolation. Assume that a destination location (x', y') maps backward to source location (x, y). The source pixel nearest location (x, y) is located at $(\mathrm{round}(x), \mathrm{round}(y))$ and the source pixel at that image is then carried over as the value of the destination. In other words, $I'(x', y') = I(\mathrm{round}(x), \mathrm{round}(y))$. Nearest neighbor interpolation is computationally efficient but of generally poor quality, producing images with jagged edges and high graininess.

Bilinear interpolation assumes that the continuous image is a linear function of spatial location. Linear, or first order, interpolation combines the four points surrounding location (x, y) according to Equation (7.8), where (x, y) is the backward mapped coordinate that is surrounded by the four known samples at coordinates (j, k), $(j, k+1)$, $(j+1, k)$ and $(j+1, k+1)$. Variable b represents the proximity of (x, y) to the lower horizontal edge and variable a represents the proximity of (x, y) to the vertical edge. Both variables a and b are weighting values:

$$
\begin{aligned}
j &= \lfloor x \rfloor, \\
k &= \lfloor y \rfloor, \\
a &= x - j, \\
b &= y - k, \\
I'(x', y') &= [1 - b, b] \cdot \begin{bmatrix} I(j, k) & I(j+1, k) \\ I(j, k+1) & I(j+1, k+1) \end{bmatrix} \cdot \begin{bmatrix} 1 - a \\ a \end{bmatrix}.
\end{aligned}
\tag{7.8}
$$

Bilinear interpolation is a weighted average where pixels closer to the backward mapped coordinate are weighted proportionally heavier than those pixels further away. Bilinear interpolation acts like something of a rigid mechanical system where rods vertically connect the four samples surrounding the backward mapped coordinate, as shown in Figure 7.5. A third rod is connected horizontally which is allowed to slide vertically up and down the fixture. In addition, a ball is attached to this horizontal rod and is allowed to slide freely back and forth across the central rod. The height of the ball determines the interpolated sample value wherever the ball is located. In this way it should be clear that all points within the rectangular area bounded by the four corner posts have implicit, or interpolated, sample values.

Higher-order interpolations may yield improvements over bilinear interpolation. Bicubic interpolation, for example, estimates values by computing a weighted sum of the 16 samples closest to the backward mapped coordinate. While bicubic interpolation generally produces better results than bilinear interpolation, it is obviously slower in terms of computation. For most practical

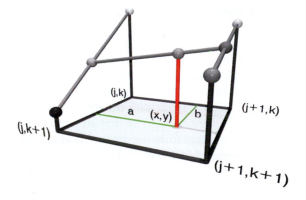

Figure 7.5. Bilinear interpolation.

Figure 7.6. Interpolation example.

purposes bilinear interpolation is sufficiently accurate and efficient. Figure 7.6 compares the effects of nearest neighbor and bilinear interpolation when rotating an image. In this example an image is rotated by $45°$ clockwise using both nearest neighbor and bilinear interpolation. The detail shown in the upper right uses nearest neighbor interpolation; the detail at the lower right uses bilinear.

7.1.4 Resampling

Interpolation is closely related to image scaling, also known as resampling, which changes the dimensions of an image by either increasing or decreasing the width and/or height of an image. Figure 7.7 shows a schematic of this process. In this figure an image is either in the continuous domain, where light intensity is defined at every point in some projection or in the discrete domain, where intensity is defined only at a discretely sampled set of points. When an image is acquired, an image is taken from the continuous into the discrete domain at which time the dimensions of the image are fixed. The first step of resampling, known as reconstruction, is to take an image from the discrete domain into the continuous domain where it can be resampled at any desired resolution. When an image is reconstructed, it is taken from the discrete into the continuous domain by using interpolation to approximate the sample values at every location in space. The reconstructed image can then be resampled to any desired resolution through the typical process of sampling and quantization.

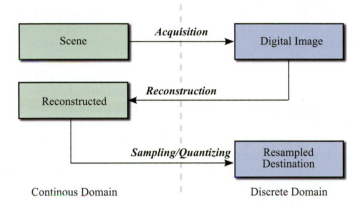

Figure 7.7. Resampling.

7.1.5   Implementation of Geometric Operations

The standard Java distribution provides an `AffineTransform` class for representing homogeneous affine transformation matrices. `AffineTransform` contains several convenient methods for creating homogeneous matrices of which a partial listing is given in Figure 7.8.

The default constructor produces an identity that which maps all locations to themselves. The second constructor of Figure 7.8 accepts a six-element array of coefficients representing the six configurable weights in the transformation matrix. This constructor gives clients the freedom to set the weights in any way they choose. The four static factory methods give clients convenient ways of constructing matrices that perform one of the four basic transforms without having to compute, on their own, the matrix coefficients. In order to construct a rotation matrix, for example, a client may use the `getRotateInstance` method

AffineTransform
`public AffineTransform()`   `public AffineTransform(double[] matrix)`

`public static AffineTranform getRotateInstance(double theta)`
 `public static AffineTranform getScaleInstance(double sx, double sy)`
 `public static AffineTranform getShearInstance(double shx, double shy)`
 `public static AffineTranform getTranslateInstance(double dx, double dy)`
 `public Point2D transform(Point2D src, Point2D dest)`
 `public void concatenate(AffineTransform xfrm)`
 `public void preConcatenate(AffineTransform xfrm)`

Figure 7.8. `AffineTransform` methods (partial listing).

and provide the desired rotation angle (in radians). The six coefficients will be automatically computed and the corresponding transformation matrix will be constructed and returned.

The `Point2D` class is defined in the `java.awt.geom` package and represents, as the name implies, a two-dimensional point. The `transform` method accepts a `Point2D` object and maps it in accord with the transformation matrix to obtain an output coordinate. If the destination point is non-null then the X and Y coordinates of the destination are properly set and the destination point is returned. If the destination point is null, a new `Point2D` object is constructed with the appropriate coordinate values and that object is returned.

The `AffineTransform` class contains a `createInverse` method that creates and returns the inverse `AffineTransform` of the caller. This method throws an exception if the matrix is not invertible. As mentioned earlier, the transformation matrix must be invertible in order to perform a geometric operation that uses backward mapping.

The `concatenate` and `preConcatenate` perform matrix multiplication. Let A be the calling `AffineTransform` object and B the `AffineTransform` that is passed as the single required parameter. The concatenate method modifies A so that A becomes AB while preconcatenation modifies A such that A becomes BA.

The `AffineTransformOp` is a `BufferedImageOp` implementation that applies an affine transformation to a source image. The `AffineTransformOp` contains an `AffineTransform`, which is used to map source to destination coordinates. An abbreviated UML class diagram of the `AffineTransformOp` is given in Figure 7.9.

Clients must construct an `AffineTransform` which is passed as a parameter to the constructor. The transformation matrix must be invertible since backwards mapping will be used to map destination locations into the source. Clients must also specify an interpolation technique by providing an integer valued constant that is either TYPE_NEAREST_NEIGHBOR, TYPE_BILINEAR, or TYPE_BICUBIC. Listing 7.1 shows how to rotate an image clockwise $45°$ using these two classes.

The `AffineTransformOp` is implemented in such a way as to extend the bounds of the destination in only the positive x and y directions while all pixels mapping to negative coordinates are ignored. The `AffineTransformOp` must

≪ *implements*   *BufferedImageOp* ≫
AffineTransformOp
`public AffineTransformOp(AffineTransform xfrm, int interpolationType)` `public AffineTransform getTransform()` `public int getInterpolationType()`

Figure 7.9. `AffineTransformOp` methods (partial listing).

```
1  BufferedImage  src  =  ImageIO.read(new  File("filename.png"));
2  AffineTransform  xfrm  =  AffineTransform.getRotateInstance(−Math.PI/4.0);
3  AffineTransformOp  op  =
4    new  AffineTransformOp(xfrm,  AffineTransformOp.TYPE_BICUBIC);
5  BufferedImage  dest  =  op.filter(src,  null);
```

Listing 7.1. Rotation with `AffineTransformOp`.

therefore be used with caution since it is possible to rotate *all* pixels onto negative coordinate values and hence to produce an *empty* destination. If, for example, an image is rotated by $180°$, all pixels except the origin fall outside of the bounds of the source. In this case the `AffineTransformOp` actually gives rise to an exception since it computes the width and height of the destination to be zero.

Listing 7.2 shows how to rotate an image such that the entire image is always displayed in the destination regardless of whether pixels are mapped into negative coordinates. In this example, line 3 constructs an `AffineTransformation` that rotates the source by $-45°$ about the origin. Since portions of the image will be rotated into quadrants with negative coordinates, lines 5–8 compute offsets to shift the image back onto the positive quadrant. Line 6 obtains the bounds of the source image as a `Rectangle` object, which is then rotated using the `AffineTransform`. The bounds of the rotated rectangle are then obtained as another `Rectangle` object. This rectangle is positioned at coordinate (destBounds.getX(), destBounds.getY()), the amount by which the rotated image must be translated. A translation matrix is constructed and pre-concatenated with the rotation matrix to obtain an `AffineTransform` that both rotates and centers a source image onto the destination.

```
1  BufferedImage  src  =  ImageIO.read(new  File("filename.png"));
2
3  AffineTransform  xfrm  =  AffineTransform.getRotateInstance(−Math.PI/4.0);
4
5  Rectangle2D  destBounds  =
6    xfrm.createTransformedShape(src.getRaster().getBounds()).getBounds2D();
7  AffineTransform  translate  =
8    AffineTransform.getTranslateInstance(−destBounds.getX(),
9                                          −destBounds.getY());
10  xfrm.preConcatenate(translate);
11
12  AffineTransformOp  op  =  new  AffineTransformOp(xfrm,  null);
13  BufferedImage  dest  =  op.filter(src,  null);
```

Listing 7.2. Rotation with `AffineTransformOp`.

```
1  BufferedImage src = ImageIO.read(new File("filename.png"));
2  AffineTransform xfrm = AffineTransform.getScaleInstance(1.5, 2.3);
3  AffineTransformOp op =
4      new AffineTransformOp(xfrm, AffineTransformOp.TYPE_BICUBIC);
5  BufferedImage dest = op.filter(src, null);
```

Listing 7.3. Scaling with `AffineTransformOp`.

When using Java's `AffineTransformOp`, reconstruction is accomplished through either nearest neighbor, bilinear or bicubic interpolation. Listing 7.3 shows how to apply non uniform scaling transformation of 1.5 horizontally and 2.3 vertically. Listing 7.3 is identical to that of 7.1 except for line 2 where an affine transform that scales an image is generated.

7.2   Custom Implementation

Implementation of a flexible image processing library requires principled object-oriented design techniques where core methods are abstracted and class coupling is minimized. Since interpolation is central to geometric transforms, an `Interpolant` class is defined that captures the central notion of interpolation and decouples interpolation from the transformation process. The `Interpolant` of Listing 7.4 is an interface containing a single method named `interpolate`. This method takes a real-valued `Point2D` and produces the sample value at that location in the source image. In addition, an `ImagePadder` object must be supplied so that the source image is extended across all space in order to allow interpolation at all spatial coordinates. Interpolation works on a single band and hence the band is also given as the final argument.

Implementations of this interface are free to define the `interpolate` method using a nearest neighbor, bilinear, bicubic technique, or any technique of their

```
1  public interface Interpolant {
2      public int interpolate(BufferedImage src,
3                             ImagePadder padder,
4                             Point2D point,
5                             int band);
6  }
```

Listing 7.4. Interpolant.

```
1  class NearestNeighborInterpolant implements Interpolant {
2    public int interpolate(BufferedImage src,
3                           ImagePadder padder,
4                           Point2D point,
5                           int band) {
6      int x=(int)Math.round(point.getX());
7      int y=(int)Math.round(point.getY());
8
9      return padder.getSample(src, x, y, band);
10   }
11 }
```

Listing 7.5. NearestNeighborInterpolant.

choosing. Listing 7.5 gives an implementation of the simplest interpolation scheme: the nearest neighbor interpolation. The NearestNeighborInterpolant finds the nearest integer coordinate via the round function and returns the corresponding source sample value provided by the image sampler.

Since geometric transformations key on a mapping between source and destination, the responsibility for mapping is abstracted into its own class. Listing 7.6 describes the InverseMapper class that decouples the task of mapping from other classes.

The most important method of the InverseMapper is the overloaded inverse Transform. This method accepts a coordinate in the *destination* image and maps

```
1  public abstract class InverseMapper {
2    public Point2D inverseTransform(int dstX, int dstY, Point2D srcPt) {
3      return inverseTransform(new Point2D.Double(dstX, dstY), srcPt);
4    }
5
6    public ImagePadder getDefaultPadder() {
7      return ZeroPadder.getInstance();
8    }
9
10   public void initializeMapping(BufferedImage src) { }
11
12   public Rectangle getDestinationBounds(BufferedImage src) {
13     return src.getRaster().getBounds();
14   }
15
16   public abstract Point2D inverseTransform(Point2D destPt, Point2D srcPt);
17 }
```

Listing 7.6. InverseMapper.

that coordinate back onto the source image. The destination coordinate is given either as a pair of integers or as a `Point2D` while the corresponding source coordinate is given as a `Point2D`. If the `srcPt` parameter is non-null, that object is filled in by the `inverseTransform` method and returned. If the `srcPt` parameter is null then the `inverseTransform` method creates a new `Point2D` object, which is returned as the result.

Some mappers may need to know something about the source image that is being transformed and hence the initialize mapping method is given. This method should always be called once prior to asking the mapper to transform coordinates. The mapper can then refer to the source image for any information that it requires. Mappers may also compute the preferred bounds of the destination given a source image. The bounding rectangle is computed by the `getDestination` method, which defaults to the same size as the source. Specific subclasses are expected to override this method if the destination bounds are to be different than the source.

Affine transformations can be easily incorporated into this framework by writing a subclass of `InverseMapper` that uses an `AffineTransform` object to perform the inverse transformation. Listing 7.7 gives a complete implementation of `InverseMapper` that relies on an `AffineTransform` object for completion of the `inverseTransform` method.

Any geometric transformation can then be implemented in terms of interpolants and mappers, as shown in Listing 7.8. The `GeometricTransformOp`

```
1  public class AffineMapper extends InverseMapper {
2    private AffineTransform inverseXfrm, forwardXfrm;
3
4    public AffineMapper(AffineTransform xfrm)
5            throws NoninvertibleTransformException {
6      forwardXfrm = xfrm;
7      inverseXfrm = xfrm.createInverse();
8    }
9
10   public Point2D inverseTransform(Point2D dstPoint, Point2D srcPt) {
11     return inverseXfrm.transform(dstPoint, srcPt);
12   }
13
14   public Rectangle getDestinationBounds(BufferedImage src) {
15     return forwardXfrm.createTransformedShape(
16                     src.getRaster().getBounds()).getBounds();
17   }
18 }
```

Listing 7.7. `AffineTransformMapper`.

is a BufferedImageOp that performs backward mapping through the inverse Transform method of a client supplied InverseMapper and interpolation through the interpolate method of a client supplied Interoplant. This design is exceptionally flexible since it is not dependent upon either a particular interpolation technique or a particular geometric transform.

Note that this class overrides the createCompatibleDestImage method such that the destination dimensions are determined by the bounds of the transformed

```java
public class GeometricTransformOp extends NullOp {
  protected Interpolant interpolant;
  protected InverseMapper mapper;
  protected ImagePadder handler;

  public GeometricTransformOp(InverseMapper mapper, Interpolant interpolant) {
    this(mapper, interpolant, mapper.getDefaultPadder());
  }

  public GeometricTransformOp(InverseMapper mapper,
                              Interpolant interpolant,
                              ImagePadder handler) {
    this.interpolant = interpolant;
    this.mapper = mapper;
    this.handler = handler;
  }

  public BufferedImage createCompatibleDestImage(BufferedImage src,
                                                 ColorModel destCM) {
    Rectangle bounds = mapper.getDestinationBounds(src);

    return new BufferedImage(destCM,
        destCM.createCompatibleWritableRaster((int)(bounds.getWidth()),
                                              (int)(bounds.getHeight())),
        destCM.isAlphaPremultiplied(),
        null);
  }

  @Override
  public BufferedImage filter(BufferedImage src, BufferedImage dest) {
    if (dest == null) {
      dest = createCompatibleDestImage(src, src.getColorModel());
    }

    Rectangle bounds = mapper.getDestinationBounds(src);
    mapper.initializeMapping(src);

    Point2D dstPt = new Point2D.Double();
    Point2D srcPt = new Point2D.Double();
    for (Location pt : new RasterScanner(dest, false)) {
      dstPt.setLocation(pt.col + bounds.x, pt.row + bounds.y);
      mapper.inverseTransform(dstPt, srcPt);
```

```
43        for (int b = 0; b < src.getRaster().getNumBands(); b++) {
44          int sample = interpolant.interpolate(src, handler, srcPt, b);
45          dest.getRaster().setSample(pt.col,
46                                     pt.row,
47                                     b,
48                sample);
49        }
50      }
51    return dest;
52  }
53 }
```

Listing 7.8. `GeometricTransformOp`.

source rather than the dimensions of the source itself. The mapper is responsible for determining the destination bounds as can be seen in line 35. When filtering any source the mapper is initialized by the call of line 36 to the initialization method. Every sample of the destination image rather than the source is then scanned as seen in line 41, and an interpolant is used to find the source sample corresponding to each destination location. Note also that the image padder is not used directly in this code but that it is handed to the interpolant for its use in line 44.

Using this framework, clients are responsible for constructing an interpolant and a mapper and applying the transform through the `GeometricTransformOp` class. A code fragment is shown in Listing 7.9, where nearest neighbor interpolation is used with a rotation mapper to perform the desired filtering operation.

Our custom design is captured in the UML diagram of Figure 7.10, where three classes collaborate to perform geometric transformations. In this diagram, we indicate the likelihood that clients will design `Interpolant`s such as the `NearestNeighbor`, `Bilinear` and `Bicubic` method in addition to others. Clients are also able to extend the `InverseMapper` so that effects like twirling, rippling, or any other mapping of their choice is supported. These nonlinear transformations are described in the following section.

```
1 BufferedImage src = ImageIO.read(new File(fileName));
2 Interpolant interpolant = new NearestNeighborInterpolant();
3 InverseMapper mapper =
4    new AffineTransformMapper(AffineTransform.getRotateInstance(Math.PI/4));
5 BufferedImageOp op = new GeometricTransformOp(mapper, interpolant);
6 BufferedImage dest = op.filter(src, null);
```

Listing 7.9. Applying a geometric transform.

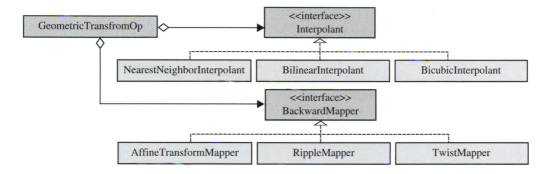

Figure 7.10. UML class diagram of the GeometricTransformOp.

7.3   Nonlinear Transformations

While linear transformations can be expressed as a 3×3 affine transformation matrix, nonlinear transformations are arbitrarily complex user-defined functions. Figure 7.11 gives a visual example of a nonlinear warp. In this example the

(a) Source image.   (b) Twirled.

Figure 7.11. Nonlinear transformation.

image is twisted or twirled about some point of reference by an arbitrary amount. This spatial warping is indicated by the superimposed grids on both the source and destination images of Figure 7.11, where the reference point is given as the center of the image.

The inverse transformation functions used to generate the twirled destination of this example are nonlinear and given by Equation (7.9), where x and y are given as offsets from a reference point defined as (centerX, centerY). The strength parameters give control over the amount of the vortex, where greater strength values indicate greater distortion and smaller strength values

```java
public class TwirlMapper extends InverseMapper {
  private double xc, yc, minDim, strength;
  private int centerX, centerY;

  public TwirlMapper(double xc, double yc, double strength) {
    this.xc = xc;
    this.yc = yc;
    this.strength = strength;
  }

  public void initializeMapping(BufferedImage src) {
    centerX = (int)(xc * src.getWidth());
    centerY = (int)(yc * src.getHeight());
    minDim = Math.min(src.getWidth(), src.getHeight());
  }

  public Point2D inverseTransform(Point2D dstPt, Point2D srcPt) {
    if(srcPt == null) {
      srcPt = new Point2D.Double();
    }
    double dx = dstPt.getX() - centerX;
    double dy = dstPt.getY() - centerY;
    double r = Math.sqrt(dx*dx + dy*dy);
    double theta = Math.atan2(dy,dx);

    double srcX = r * Math.cos(theta+strength*(r-minDim)/minDim) + centerX;
    double srcY = r * Math.sin(theta+strength*(r-minDim)/minDim) + centerY;

    srcPt.setLocation(srcX, srcY);
    return srcPt;
  }
}
```

Listing 7.10. TwirlMapper.

(a) Warped.   (b) Rippled.   (c) Diffused.

Figure 7.12. Nonlinear geometric transformations.

produce less distortion:

$$T_x(x,y) = r \cdot \cos(\theta + \text{strength} \cdot (r - \text{minDim})/\text{minDim}) + \text{centerX},$$
$$T_y(x,y) = r \cdot \sin(\theta + \text{strength} \cdot (r - \text{minDim})/\text{minDim}) + \text{centerY}. \tag{7.9}$$

This nonlinear mapping can be integrated into our custom geometric operation framework simply be subclassing the inverse mapper. Consider, for example, Listing 7.10 , which is a complete implementation of a InverseMapper that achieves the effects shown in Figure 7.11. When constructing a TwirlMapper object, clients are able to specify the amount of distortion by specifying the strength value. In addition, the distortions are made with respect to a reference point, which is given as the first two constructor parameters xc and yc specifying, in this listing, the relative location of the reference point. In other words, parameters xc and yc are fractions of the width and height of the source, respectively, such that to specify the center of any source image the parameters should both be set to 0.5.

Other effects can be relatively easily implemented and are left as exercises. Figure 7.12 gives an example of nonlinear geometric transformations. The image of part (a) is warped such that space closer to the center is expanded while space closer to the periphery is compressed. The image of part (b) is rippled both vertically and horizontally while the image of (c) is diffused. Under diffusion, a point is randomly shifted by an arbitrary amount from it's original location.

Pixel Jelly fully supports the geometric transforms described in this section. In addition, effects such as a fisheye lens and a kaleidoscope effect are also included. Figure 7.13 shows a screenshot of the geometric operations dialog window.

Figure 7.13. Geometric Ops screenshot.

7.4 Exercises

1. Give the homogeneous transformation matrix representing the sequence of transformations given by (1) translation by $(-50, 50)$ followed by (2) rotation around the origin by $45°$ followed by (3) a horizontal shear with a shear factor of 1.25.

2. Transform the point $(25, 30)$ using the homogeneous transformation matrix below:
$$\begin{bmatrix} 1.59 & 0.18 & -70.7 \\ 0.71 & 0.71 & 0.0 \\ 0.0 & 0.0 & 1.0 \end{bmatrix}.$$

3. Design and implement a class named `ForwardGeometricOp` that performs forward mapping. The class should use a mapper in a way that is similar to the custom `GeometricOp` technique mentioned in the text. The mapper for this class should map source coordinates into integer-valued destination coordinates. No interpolation is required in this operation since each sample of the source is simply copied to an appropriate location in the destination.

4. Design and implement a class named `BilinearInterpolant` that implements the `Interpolant` interface of Listing 7.4. Your implementation should, of course, perform bilinear interpolation.

5. Design and implement a class named `DecimationOp` that extends `NullOp`. Decimation is a technique for reducing the size of an image by simply deleting every ith row and every jth column where i and j are controlling parameters. When clients construct a `DecimationOp` they will specify the values of i and j. Execute your filter on a variety of image types and discuss why this technique is either superior or inferior to the resampling technique described in this text.

6. Design and implement a class `DiffuseMapper` that extends the `Inverse Mapper` class of Listing 7.6. The `DiffuseMapper` should produce images similar to that shown in Figure 7.12 (c). The mapping between destination and source is given by

$$
\begin{aligned}
x &= x' + S \cdot rand(), \\
y &= y' + S \cdot rand().
\end{aligned}
$$

where S is a parameter giving the strength of the diffusion and $rand()$ is a function returning a uniformly distributed random value in the interval $[-1, 1]$.

★ 7. Design and implement a class `RippleMapper` that implements the `InverseMapper` interface of Listing 7.6. The `RippleMapper` produces images that are rippled in both the horizontal and vertical dimension about a reference point (xc, yc) with strengths S_x and S_y. Letting $dx = x' - xc$, $dy = y' - yc$, $r = \sqrt{dx^2 + dy^2}$, $\theta = \tan^{-1}(dy, dx)$, and $D = \min(\text{width}, \text{height})$; then the twisting transformation is given by

$$
\begin{aligned}
x &= r \cdot \cos(\theta + S_x \cdot (dy - D)/D) + xc, \\
y &= r \cdot \sin(\theta + S_y \cdot (dx - D)/D) + yc.
\end{aligned}
$$

Artwork

Figure 7.1. *Portrait of Victorine Meurent* by Édouard Manet (1832–1883). See page 253 for a brief biographical sketch of the artist. Although Victorine Meurent was an able artist, she is best remembered as a model for Édouard Manet and appears in his two masterworks, *The Luncheon on the Grass* and *Olympia*. She began modeling at the age of sixteen and was inducted into the Société des Artistes Français in 1903.

Figure 7.2. *Portrait of Victorine Meurent* by Édouard Manet (1832–1883). See page 253 for a brief biographical sketch of the artist and the previous entry for a brief discussion of Victorine Meurent.

Figure 7.3. *The Birth of Venus* by Allessandro di Mariano di Vanni Filipepi (1445–1510). Alessandro Filipepi was an Italian painter during the Early Renaissance period and is more commonly known as Sandro Botticelli. Figures 7.3 and 7.6 depict a detail of his most famous work showing the goddess Venus emerging from the sea as a full grown adult figure.

Figure 7.6. *The Birth of Venus* by Allessandro di Mariano di Vanni Filipepi (1445–1510). See previous entry for a brief biographical sketch of the artist.

Figure 7.11. *Girl with Golden Hat* by Pierre-August Renoir (1841–1919). See page 119 for a brief biographical sketch of the artist.

Figure 7.12. *Girl with Golden Hat* by Pierre-August Renoir (1841–1919). See page 119 for a brief biographical sketch of the artist.

Image Printing and Display

8

Digital images are meant to be seen and therefore displayed on some visual medium. A digital image may be displayed on a computer monitor or television screen, printed in a newspaper or produced on a large variety of output devices each of varying capabilities. Processing is usually required to accommodate the properties of the output device so that the displayed image appears identical to its corresponding source image. This processing generally attempts to mimic the color depth, color model, and spatial resolution of the source image where the output device may have a reduced color depth, have a lower resolution, or doesn't support the same color model.

An output device will generally be characterized by pixel shape, spatial resolution, and color depth. A typical computer monitor, for example, will use square pixels with a spatial resolution of 72 pixels per inch and a color depth of 32 bpp. A black-and-white laser printer, by contrast, may use circular pixels with a resolution of 1200 pixels per inch and a color depth of 1 bpp. Whenever a digital image is rendered for display, the characteristics and limitations of the output device must be considered in order to generate an image of sufficient fidelity.

8.1 Halftoning

The central problem when printing relates to the color depth of the output device. Many printers support only a 1 bpp color depth since they only use black ink placed on white paper. This naturally raises the question of how a grayscale image can be produced on a binary output device. In the case of a binary output, halftoning seeks to answer the question of which output pixels in the output should be black and which white in order to most accurately reproduce a grayscale source. The central question then becomes how to reproduce a grayscale image on an output device that can only print with black ink. In more formal terms, the question becomes that of reducing the color depth from

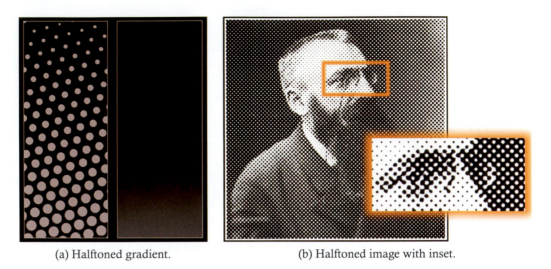

(a) Halftoned gradient.   (b) Halftoned image with inset.

Figure 8.1. Traditional halftoning.

an 8 bpp grayscale source image to a 1 bpp destination while maintaining an acceptable visual appearance.

In traditional, analog halftoning, a grayscale image is converted into a binary image composed of a pattern of dots. The dots are arranged in a grid and are themselves of various sizes. Figure 8.1 shows how black dots of various sizes printed on a white background can give the visual illusion of all shades of gray when viewed from an appropriate distance. The 8-bit grayscale gradient of (a) is halftoned to a 1-bit approximation. While appearing to be a grayscale image, the image of part (b) is a 1 bpp halftone as depicted by the highlighted inset. Halftoning in this example gives a 1-bit output device the illusion of being an 8-bit device.[1]

Traditional halftoning is a continuous domain or analog process that is performed by projecting an image through an optical screen onto film. The surface of the optical screen is etched with lines in such a way as to cause dots to appear on the film in correspondence with the source intensity. Larger black dots appear in regions of dark intensity and smaller black dots in regions of bright intensity. The spatial resolution of halftone systems is given as lines per inch (LPI), which measures the density of the etched lines on the optical screen. Newsprint, for example, is typically printed at 85 LPI while glossy magazines are printed using 300 LPI halftone screens.

---

[1]It should be noted that since this text is printed with black ink on white background all images included herein have been halftoned. Even those images that appear grayscale in the text are, of course, binary.

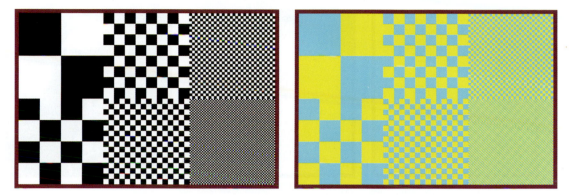

Figure 8.2. Dithering example.

Digital halftoning, also known as dithering, is any binary process that reduces the color depth of a source while maintaining the sources spatial resolution. In image processing, a binary process is any process that outputs one of two colors for every pixel. Digital halftoning differs from traditional halftoning since the digital halftoning process is discrete and the spatial resolution of the source image and the output device are uniform. Traditional halftoning takes place in the continuous domain and the spatial resolution of the output device is flexible since dot sizes are allowed to vary continuously. The output device normally has a color depth of 1 bpp; hence the task is to convert a grayscale image into a binary image of the same dimensions.

The human visual system performs spatial integration (averaging) of colors near the point of focus and hence the central idea of dithering is to ensure that the local average of all output samples is identical to its corresponding source sample. Dithering increases the apparent color depth of an output device by carefully intermingling colors from some limited palette in such a way that when local regions are averaged they produce the desired colors. Color dithering is discussed later in this chapter. Figure 8.2 illustrates how various shades of gray can be generated by interweaving only the colors black and white, as shown in the leftmost image. Figure 8.2 also shows how shades of green can be generated by interweaving only cyan and yellow.

8.2 Thresholding

Thresholding is a point processing technique that is also one of the simplest dithering techniques. For a given source sample S_{input} and threshold value τ, the corresponding output sample S_{output} is given in Equation (8.1). The output is either black or white, depending upon whether or not the input sample is below

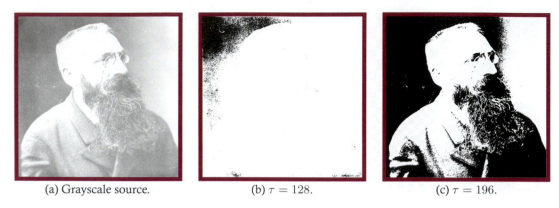

(a) Grayscale source.   (b) $\tau = 128$.   (c) $\tau = 196$.

Figure 8.3. Threshold selection.

or above the threshold:

$$S_{\text{output}} = \left\{ \begin{array}{ll} \text{black} & S_{\text{input}} < \tau, \\ \text{white} & S_{\text{input}} \geq \tau. \end{array} \right. \qquad (8.1)$$

Thresholding rarely produces pleasing results and is solely dependent on proper selection of the threshold value. The threshold is commonly set to the center of the source image's dynamic range, which for an 8-bit image equates to 128. While this threshold value is appropriate as a generic solution it does not produce good results in many cases. Figure 8.3 illustrates the effect of choosing an incorrect threshold. In this example, an overexposed source image is thresholded with a cutoff of 128 to obtain the binary image of (b). Since the source image was overexposed, nearly all of the grayscale values in the source exceed 128 and hence nearly all of the resulting binary output samples are converted to white. Choosing a threshold of 196 produces much better results as can be seen in Figure 8.3(c).

Adaptive thresholding, also known as dynamic thresholding, is used to determine an appropriate threshold for a particular image. Adaptive thresholding is typically based on a statistical analysis of an image's histogram, and seeks to determine an optimal split between clusters of samples in the data distribution. The simplest adaptive thresholding technique is to use either the average or median value of all source samples as the threshold. Computing both the average and mean sample values requires one pass through the image data and hence incurs a small amount of overhead.

A more sophisticated alternative is to use is an iterative technique uncovered by Ridler and Calvard [Ridler and Calvard 78]. This algorithm locates a threshold that is midway between the means of the black and white samples in the histogram. Listing 8.1 gives a pseudocode description of the algorithm that is

```
1  algorithm findThreshold(Image src) {
2      T0 = 0
3      T1 = 128
4      while(T0 != T1)
5          T0 = T1
6          M0 = the average value of all samples < T0
7          M1 = the average value of all samples >= T0
8          T1 = (M0+M1)/2
9      return T0
10 }
```

Listing 8.1. Adaptive threshold selection.

straightforward to implement but is computationally expensive since it converges to a solution by making multiple passes through the image.

Regardless of the technique used to compute a threshold value, adaptive thresholding can be either global or local. Global adaptive thresholding uses a single, fixed threshold value that is applied to every sample in the source while local adaptive thresholding uses a separate threshold value for each sample in the source image. Local thresholding examines the samples in a region centered on an input sample and computes a threshold value that is used only for that particular sample.

Local adaptive thresholding can be described as a sequence of other well-known processes. Let I be a source image that is to be locally thresholded and let M be the result of median filtering I. Image M represents the threshold values for each sample of I such that every sample $I(x, y, b)$ that exceeds $M(x, y, b)$ should be white. Image M is then subtracted from I, which results in an image having positive samples wherever $I(x, y, b) > M(x, y, b)$ and zero-valued samples elsewhere (assuming that clamping is used when subtracting). This image can then be globally thresholded using a fixed threshold value of 1.

Local adaptive thresholding is useful as a preprocessing step in certain computer vision applications that attempt to identify objects in an image with varying illumination. Figure 8.4 shows an 8 bit grayscale source image that has been globally thresholded using an adaptively determined threshold value in part (b) and one where the threshold is adaptively determined based on the local region, as shown in (c).

As mentioned earlier in this section, the goal of halftoning is to produce an output where the average light intensity of every local region in the output image is approximately equal to the corresponding average in the same region of the source. Although thresholding achieves such an equivalence on a global scale it rarely achieves this effect on smaller local scale. Consider again the images of Figure 8.4. The grayscale source has an average value of 88 while the binarized

(a) Source image.   (b) Global.   (c) Local.

Figure 8.4. Thresholding.

image of (c) has an average value of 65. This is a reasonable approximation of grayscale averages over the entire image. Consider, however, the local region of the upper-right shoulder where the binary image has an average grayscale value of 255 while the corresponding image has a grayscale average of approximately 180, depending upon precise location. In other words, while prominent features of an image survive the thresholding process, much of the detail is obliterated.

8.3   Patterning

Dithering techniques are a significant improvement over thresholding since they generate images that maintain approximately equal intensities at the local level and not only at the global level. *Patterning* is a halftone dithering technique that generates a binary image from a grayscale or color source by increasing the resolution of the output in order to compensate for the decreased color depth. Patterning works by using a group of pixels in the display device to represent a single pixel from the source image.

Consider using a 3×3 binary region to represent a single grayscale value. There are 512 possible patterns or arrangements of the nine binary pixels, but fewer possible *average intensity values*. With nine binary pixels, a 3×3 region is able to promulgate the illusion of only ten shades of gray. When all pixels are black, the average value of the nine pixels is $0/9$. When any one pixel is white and all others are black, the average value is $1/9$. This logic is easily extended to show that in normalized grayscale color space a 3×3 binary image is able to simulate ten shades of gray having normalized values of $\frac{1}{9}\{0, 1, 2, 3, 4, 5, 6, 7, 9\}$. A binary font is a set patterns such that all possible grayscale values are represented. Figure 8.5 shows an example of a 3×3 binary font.

Figure 8.5. Clustered dot patterning example.

Certain font patterns should be avoided since they will cause artificial lines or repeated patterns in the destination. Diagonal, horizontal, or vertical lines within a single pattern should generally be avoided. The font of Figure 8.6 is known as an *clustered* pattern since as intensity increases each new pattern is obtained by flipping one black dot to white. In other words, each pattern is a superset of the previous pattern in the sequence. Any halftoning technique is subject to distortion by using a device that slightly smears or blurs individual pixels when printed. The clustered dot technique minimizes the effect of such smearing since the blurry black pixels tends to overlap other blurry black pixels.

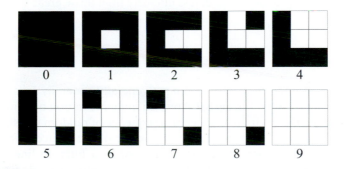

Figure 8.6. A 3×3 binary font.

Of course the destination image will be larger in size than the source image by an amount that is proportional to the color depth of the destination. As the example of Figure 8.6 indicates, an output image with 10 grayscale levels requires a threefold increase in dimension. To generate an image with 256 levels of intensity requires a large font size where the minimum size of such a font is left as an exercise for the reader. Since large font sizes are required to obtain good halftones, patterning is rarely used in practice. Figure 8.5 shows a source image that uses the clustered dot pattern to produce the binary image on the right. Notice that the destination image is three times the width and height of the source and that artificial contours are introduced as an artifact of patterning.

8.4   Random Dithering

Random dithering, as its name implies, chooses the threshold value at random from a uniform distribution of values in the dynamic range of the source. Although it may not be intuitively apparent, this technique does maintain both the global intensity value and local intensity values over reasonably small neighborhoods. Consider a grayscale image having an average grayscale intensity of 100. On average, the randomly selected threshold will fall below the pixel value approximately 100 out of every 255 samples, thus generating a white output, while about 155/255 percent of the thresholds will be above the pixel value and hence will likely generate a black output, thus maintaining the proper average intensity value at any dimensional scale.

Digital random thresholding is similar to a high quality printmaking technique known as *mezzotinting*, a process that is known to have been used since 1642. In mezzotinting, an artist roughens the surface of a soft metal printing plate with thousands of small randomly located depressions or dots. The density of the dots within a local region determines the tonality of the print. When the plate is covered with ink and pressed against canvas or paper, those regions with a high dot density produce areas of less intensity than those areas with few or no dots.

8.4.1   Implementation

While simple to implement, this technique generally generally fails to produce acceptable visual results. Listing 8.2 shows an implementation of random dithering. The Random class is used to generate a sequence of random thresholds in the range 0 to 256 which are applied to each sample of the source. Note that the brightness band of the source image is extracted by a filter on line 28 prior to thresholding. This effectively converts color images into grayscale images prior to thresholding while leaving grayscale images intact.

```
1  public class RandomDitheringOp extends NullOp {
2    private Random randomNumbers;
3
4    public RandomDitheringOp(int seed) {
5      randomNumbers = new Random(seed);
6    }
7
8    public RandomDitheringOp() {
9      randomNumbers = new Random();
10   }
11
12   @Override
13   public BufferedImage filter(BufferedImage src, BufferedImage dest) {
14     if(dest != null && dest.getType() != BufferedImage.TYPE_BYTE_BINARY) {
15       throw new IllegalArgumentException("destination must be binary");
16     }
17
18     if (dest == null) {
19       dest = new BufferedImage(src.getWidth(),
20                                src.getHeight(),
21                                BufferedImage.TYPE_BYTE_BINARY);
22     }
23
24     if(src.getType() == BufferedImage.TYPE_BYTE_BINARY) {
25       return new NullOp().filter(src, dest);
26     }
27
28     src = new BrightnessBandExtractOp().filter(src, null);
29     for (Location pt : new RasterScanner(src, false)) {
30       int threshold = randomNumbers.nextInt(257);
31       int brightness = src.getRaster().getSample(pt.col, pt.row, 0);
32       int output = 255;
33       if (brightness < threshold) {
34         output = 0;
35       }
36       dest.getRaster().setSample(pt.col, pt.row, 0, output);
37     }
38     return dest;
39   }
40 }
```

Listing 8.2. `RandomDitherOp`.

8.5   Dithering Matrices

A dither matrix is a rectangular pattern of threshold values that seeks to produce optimal output for a local region of the source. When dithering a $W \times H$ source image with a $N \times N$ dither matrix, the dithering matrix is generally much

0	8	2	10
12	4	14	6
3	11	1	9
15	7	13	5

6	7	8	9
5	0	1	10
4	3	2	11
15	14	13	12

56	72	88	104
40	8	24	167
72	56	40	183
247	231	215	199

(a) Ordered dot.　　　　(b) Clustered dot.　　　(c) Rescaled clustered dot.

Figure 8.7. Dither matrices.

smaller than the source and is therefore repetitively tiled across and down the source in order to generate threshold values for every source sample. Dither matrices correspond to pattern fonts since the thresholds generally correspond to the likelihood of a black pixel occurring in any one of the fonts. For the font of Figure 8.6 the center threshold would have a value of 0 since it is black only once of the ten patterns while the bottom right pixel has a threshold of 15 since it is the most probable black pixel within the font.

Dither matrices are generally square. Figure 8.7 shows two 4×4 dither matrices. For each matrix the thresholds cover the interval [0,15] and represent 16 different grayscale values. In order to generate actual threshold values for a particular source image, the matrix must be rescaled to the dynamic range of the source. Given an $N \times N$ dither matrix D_m we can rescale the matrix for 8 bpp images to generate the threshold matrix as

$$\text{round} \left(\frac{255}{2 \cdot N^2} (2 \cdot D_m + 1) \right).$$

8.5.1 Implementation

The DitherMatrixOp class is given in Listing 8.3. Clients construct a dithering matrix operator by passing in a non-normalized dither matrix. Two publicly predefined matrices are defined as ORDERED_3x3 and CLUSTERED_3x3.

```
1  public class DitherMatrixOp extends NullOp {
2    private int [][] thresholds;
3
4    public static final int [][] ORDERED_3x3 = {{1, 7, 4},
5                                                 {5, 8, 3},
6                                                 {6, 2, 9}};
7    public static final int [][] CLUSTERED_3x3 = {{8, 3, 4,},
8                                                   {6, 1, 2},
9                                                   {7, 5, 9}};
```

```
10
11    public DitherMatrixOp(int [][] matrix) {
12       thresholds = scale(matrix);
13    }
14
15    private int [][] scale(int [][] ditherMatrix) {
16       int size = ditherMatrix.length;
17       int [][] result = new int[size][size];
18       double factor = 1.0 / (2.0 * size * size);
19       for (int x = 0; x < size; x++) {
20          for (int y = 0; y < size; y++) {
21             result[x][y] =
22                ImagingUtilities.clamp(255.0*(2*ditherMatrix[x][y]+1)*factor, 0, 255);
23          }
24       }
25       return result;
26    }
27
28    public BufferedImage filter(BufferedImage src, BufferedImage dest) {
29       if (dest == null) {
30          dest = new BufferedImage(src.getWidth(),
31                                   src.getHeight(),
32                                   BufferedImage.TYPE_BYTE_BINARY);
33       }
34       int size = thresholds.length, threshold, output;
35
36       for (Location p : new RasterScanner(src, false)) {
37          threshold = thresholds[p.col % size][p.row % size];
38          output = 0;
39          if (src.getRaster().getSample(p.col, p.row, 0) >= threshold) {
40             output = 1;
41          }
42          dest.getRaster().setSample(p.col, p.row, 0, output);
43       }
44       return dest;
45    }
46 }
```

Listing 8.3. `DitherMatrixOp`.

8.6   Error Diffusion Dithering

Error diffusion is a type of halftoning that keeps track of the error generated when mapping a source sample onto the output device and compensating for that error when mapping future pixels. Consider an 8 bit source sample valued at 200. Using a threshold value of 128 this sample maps to white (255) on the output device. The rounding error in this example is $200 - 255 = -55$. In other words, the output at that point is too bright by 55 units, and hence 55 units worth of intensity must somehow be removed from nearby locations of the destination image. Subsequent mappings will then account for this over-brightness by making adjustments such that generating a white sample is less likely by an amount directly proportional to the accumulated rounding error. Error diffusion dithering therefore maintains the grayscale tonality at the local level in addition to ensuring that the tonality of the output image is similar to the source at the global level.

8.6.1 Floyd-Steinberg Dithering

Floyd-Steinberg error diffusion is based on a raster scan of a source image that proceeds in top-to-bottom, left-to-right fashion. Each input sample is thresholded using a fixed global threshold value and a rounding or quantization error for that sample is determined. The rounding error is then split into four pieces and each of these pieces is 'pushed' into, or diffused into, nearby samples. Figure 8.8 illustrates the diffusion process where the rounding error e of a sample is split into four segments using the proportions $\{\frac{7}{16}, \frac{3}{16}, \frac{5}{16}, \frac{1}{16}\}$, which together constitute the diffusion matrix. Nearby samples are then adjusted by adding one of the error components to the sample value in that cell. The sample to the east is increased by $\frac{7e}{16}$, the sample to the southwest is increased by $\frac{3e}{16}$, the sample directly to the south is increased by $\frac{5e}{16}$, and the sample to the south-east is increased by $\frac{e}{16}$.

A positive rounding error indicates that the output image is too dark and that it should therefore be more likely that subsequent thresholding operations will generate a white output. This is exactly what occurs since the four neighboring sample values are increased and hence more likely to exceed the threshold value. Similar logic applies with a negative rounding error. In this case, the output image is too bright and subsequent thresholding should more likely be made to output a black pixel. Since neighboring pixels in this case are reduced in magnitude, those samples are more likely to fall below the threshold value and thus generate a black output. The Floyd-Steinberg error diffusion process is described algorithmically in Listing 8.4.

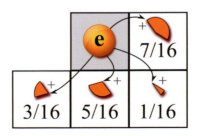

Figure 8.8. Floyd-Steinberg error diffusion.

```
1  algorithm floydSteinberg(Image src) {
2      Image temp = copy of src
3      for every sample S at location x, y
4          if(S < 128) then dest(x,y,b) = black
5          else dest(x,y,b) = white
6
7          roundingError = S − dest(x,y,b)
8          tmp(x+1,  y,b) = tmp(x+1,  y,b) + roundingError*7/16;
9          tmp(x−1,y+1,b) = tmp(x−1,y+1,b) + roundingError*3/16;
10         tmp(x,  y+1,b) = tmp(x,  y+1,b) + roundingError*5/16;
11         tmp(x+1,y+1,b) = tmp(x+1,y+1,b) + roundingError*1/16;
12     }
13 }
```

Listing 8.4. Floyd-Steinberg algorithm.

Variants of the Floyd-Steinberg diffusion matrix have since been proposed. Each of these variations seeks to improve either the aesthetic quality of the output or improve the processing speed of the filter or both. The *Jarvis, Judice, and Ninke* filter seeks to improve the quality by distributing the error across a larger neighborhood. The *Stucki* filter was later proposed as an improvement to the Jarvis-Judice-Ninke filter. The *Sierra* filter is interesting since it is both faster, in its most compact form, and typically produces more pleasing results than the Floyd-Steinberg filter. The diffusion matrices for each of these variants is shown in Figure 8.9, where the black square symbol, ■, corresponds to the currently scanned pixel and the amount of error distributed to the various neighbors is given by the matrix weights.

Figure 8.10 gives a comparison among a portion of the dithering algorithms described in this text. The image of part (a) shows the results of Floyd-Steinberg, the Jarvis-Judice-Ninke approach is shown in (b) and Sierra dithering is shown in (c).

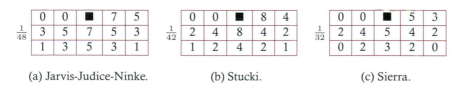

(a) Jarvis-Judice-Ninke.   (b) Stucki.   (c) Sierra.

Figure 8.9. Diffusion Matrices.

(a) Floyd-Steinberg.   (b) Jarvis-Judice-Ninke.   (c) Sierra.

Figure 8.10. Dithering example.

8.6.2   Implementation

Error diffusion dithering can be generalized as a thresholding operation followed by an error distribution step. Since the error distribution is dependent on the diffusion matrix, we choose to implement a generalized ErrorDiffusionOp class that performs the dithering while concrete subclasses simply specify the appropriate diffusion matrix. Listing 8.5 gives the implementation. Clients can construct an ErrorDiffusionOp by providing the diffusion matrix where the width of the matrix must be odd and the center element of the first row will always correspond to the scanned sample. The destination image for this operation is forced to be of binary type, although we will see later on in this chapter that diffusion can be used with color images also.

Note that within the filter method, if the source image is already binary then a copy of the source is returned by line 46. If the source image is not already

```
1  public class ErrorDiffusionBinaryOp extends NullOp {
2      private float [][] diffusionMatrix;
3      private int dmWidth,  dmHeight;
4      private ImagePadder sampler = BlackFramePadder.getInstance();
5
6      public ErrorDiffusionBinaryOp(float [][] matrix) {
7          diffusionMatrix = matrix;
8          dmWidth = diffusionMatrix[0].length;
9          dmHeight = diffusionMatrix.length;
10     }
11
12     protected void diffuseError(BufferedImage tmp,
13                                 int col,
14                                 int row,
15                                 float error) {
```

```
16
17      for (int dmRow = 0; dmRow < dmHeight; dmRow++) {
18        for (int dmCol = 0; dmCol < dmWidth; dmCol++) {
19          float weightedError = diffusionMatrix[dmRow][dmCol] * error;
20          if(weightedError != 0) {
21            float sampleWithError = sampler.getSample(tmp,
22                                              col+dmCol-dmWidth/2,
23                                              row+dmRow,0)+weightedError;
24            ImagingUtilities.safelySetSample(tmp,
25                                              col+dmCol-dmWidth/2,
26                                              row+dmRow,
27                                              0,
28                                              sampleWithError);
29          }
30        }
31      }
32    }
33
34    public BufferedImage filter(BufferedImage src, BufferedImage dest) {
35      if (dest != null && dest.getType() != BufferedImage.TYPE_BYTE_BINARY) {
36        throw new IllegalArgumentException("destination␣must␣be␣binary");
37      }
38
39      if (dest == null) {
40        dest = new BufferedImage(src.getWidth(),
41                                 src.getHeight(),
42                                 BufferedImage.TYPE_BYTE_BINARY);
43      }
44
45      if (src.getType() == BufferedImage.TYPE_BYTE_BINARY) {
46        return new NullOp().filter(src, dest);
47      }
48
49      src = new BrightnessBandExtractOp().filter(src, null);
50      src = new ConvertToFloatOp().filter(src, null);
51      for (Location pt : new RasterScanner(src, false)) {
52        float sample = src.getRaster().getSampleFloat(pt.col, pt.row, 0);
53        int output = sample < 128 ? 0 : 255;
54        float error = sample - output;
55
56        dest.getRaster().setSample(pt.col, pt.row, 0, output);
57        diffuseError(src, pt.col, pt.row, error);
58      }
59      return dest;
60    }
61  }
```

Listing 8.5. ErrorDiffusionBinaryOp.

```
1  public class FloydSteinbergDitheringOp extends ErrorDiffusionBinaryOp {
2    public static float[][] getMatrix() {
3      return new float[][]{{0f, 0f, 7f/16f},
4                           {3f/16f, 5f/16f, 1f/16f}};
5    }
6
7    public FloydSteinbergDitheringOp() {
8      super(getMatrix());
9    }
10 }
```

Listing 8.6. `FloydSteinbergDitheringOp`.

binary then the brightness band of the source image is extracted and the raster converted to floating point type by line 50. This image serves as a temporary data structure, that can be modified, as is implicitly done in line 57, without altering the source image itself. The diffusion method of line 15 iterates over all diffusion matrix coefficients and updates the corresponding sample by the weighted error. Note that the `safelySetSample` method of line 28 allows the indices of a sample to be off of the image, in which case the method has no effect.

We can now easily implement a specific dithering technique by authoring a concrete subclass that constructs the appropriate diffusion matrix. Listing 8.6 gives a complete class that performs Floyd-Steinberg dithering.

8.7   Color Dithering

Previous sections have presented the notion of halftoning grayscale images to binary images, but the problem of dithering can be extended to color, where the idea is to reduce the color depth of a true color image to match that of the output device. A color printer, for example, may use a palette of four colors (cyan, magenta, yellow, and black) or a thin-client web browser may use the web-safe palette consisting of only 216 colors to display an image.

Figure 8.11 gives an example of color dithering. The true color 24-bit source image has a total of 206,236 unique colors. The image of part (c) has been color dithered using a palette of 16 colors that were carefully chosen to maximize the quality of the dithered image. The source image was dithered using a 4-color palette to obtain the surprisingly good result shown in (e). The detailed subimage of parts (b), (d), and (f) show how the flower in the vase at the left-hand side of the table setting is affected by color dithering.

(a) A 206,236 color source.

(b) Detail.

(c) A 4 bpp source.

(d) Detail.

(e) A 2 bpp source.

(f) Detail.

Figure 8.11. Color dithering.

```
1  algorithm colorDither(Image src, Color[] palette) {
2    for every pixel at location (x,y) in src
3      index = index of palette color closest to src(x,y)
4      dest(x,y) = palette[index]
5      error = src(x,y) − dest(x,y)
6      diffuse the error to nearby pixels on a band−by−band basis
7    end
8    return dest
9  end
```

Listing 8.7. Color dithering.

Consider a source image I and a palette P that contains n colors. For every pixel $I(x, y)$ of the source, the output pixel $I'(x, y)$ must be an element of the palette. We can extend the notion of error-diffusion dithering to color images as described in Listing 8.7. For every pixel of the source image we choose the nearest color in the palette and output that color. The error between the palette color and source is then distributed across every band of unprocessed pixels.

This algorithm measures the distance between two colors by either the L_1 or L_2 metrics described in Section 3.2.6. For each source pixel $I(x, y)$ the palette color of the smallest distance from $I(x, y)$ is chosen as the output and the error is computed and distributed on each band. A palette containing only the two colors black and white will produce a binary image in the manner described earlier in this chapter. A palette containing the colors cyan, magenta, yellow and black will produce an output suitable for printing on a four color device. When developing graphics for use on the Internet, an artist might choose the web-safe palette, a standard palette that is supported by the vase majority of web browsers.

Our presentation has thus far focused on a scenario where a color image is to be displayed on some limited resource output device where the capabilities of the output device define the palette. Certain image compression techniques, however, work by dithering a source image using a palette that is dynamically created and optimized for the source image. GIF compression, for example, takes a 32 bpp source image and dithers it according to an arbitrary palette to produce an 8-bit indexed image.

Since the fidelity of the compressed image depends on the palette, it is reasonable to choose a palette that will produce the highest quality output. In GIF compression the problem is first to find the palette that ensures the highest fidelity after dithering. This problem is discussed in more detail in Section 10.6.

8.8   Exercises

1. Use the Pixel Jelly application to generate various 3×3 fonts that are used to binarize a test image with some large low frequency regions. One of your fonts should include a horizontal row of black pixels, another should include a vertical column of black pixels, and another should have a diagonal line. Submit a report that includes the test image, each pattern used, and the results of patterning; comment on the results of binarization using these various patterns.

2. Use the Pixel Jelly application to dither a full-color source image using a palette of exactly six colors that you believe will produce the best results. Have the application choose a palette of size six that is used to dither the same source. Submit a report that includes the test image, a graphic of the palettes used, and the results of dithering. Comment on the results.

3. Write two complete subclasses of the `ErrorDiffusionBinaryOp` that use the Sierra matrix and the Jarvis-Judice-Ninke matrix for diffusion.

4. Write a complete subclass of the `ErrorDiffusionBinaryOp` that uses the Sierra-2-4A filter shown below. This filter is exceptionally small, and hence efficient. Compare the results produced by this filter with the Floyd-Steinberg and Stucki filters on a variety of image types.

5. When using patterning to convert a grayscale image to binary, a 3×3 font is required to achieve the appearance of ten grayscale levels in the output. What size font is required to achieve the illusion of 256 levels?

6. Here is an interesting way of creating an image that can be displayed using a *text editor*! Write a class that uses patterning to generate a text file such that each character of the text file corresponds to a single pixel of the source image. You might consider using a space character for white pixels and an ampersand for black pixels. Your class should take a `BufferedImage` and produce a text file. For further work consider writing a class that reads a text file and creates a `BufferedImage`. To view the "image" simply open the text file with any text editor. You may need to turn off wrapping and reduce the font size to produce an acceptable result.

Artwork

Figure 8.1. "Auguste Rodin" by Gaspar-Félix Tournachon (1820–1910). See page 119 for a brief biographical sketch of the artist. Auguste Rodin was a French sculptor whose work emphasized physical realism over the traditional stylized forms and themes. Rodin was born in Paris to a working-class family and was largely self taught; taking an interest in clay sculpture at an early age. Although controversy surrounded him throughout his career, he was, at the end of his life, recognized as the preeminent French sculptor of his day. His best known works include *The Thinker* and *The Walking Man*.

Figures 8.3, 8.4, 8.5, and 8.10. "Auguste Rodin" by Gaspar-Félix Tournachon (1820–1910). See above.

Figure 8.11. *Still Life with Hyacinth* by August Macke (1887–1914). August Macke was a German painter at the turn of the 20th century. Macke was a member of the Blue Riders, a group of Expressionist artists. Macke was acquainted with artists such as Franz Marc, Wassily Kandinsky, Robert Delaunay, and Paul Klee, whose work profoundly influenced his own style. Macke died as a solider during the early months of World War 1.

Frequency
Domain

9

9.1 Overview

Recall that a digital image is a discrete approximation to a continuous function. The digital image is acquired by sampling and quantizing some continuous image and has so far been represented as a tabular arrangement of samples. This representation lies within the spatial domain, but other representations are possible and, in certain situations, advantageous.

An alternative to representing image data spatially is to use a frequency-based approach where an image is decomposed into components that describe the frequencies present in the image. Just as a musical chord can be described as the sum of distinct pure tones, a digital image can be described as the sum of distinct patterns of varying frequencies. When an image is described in terms of its frequencies, the image is said to be represented in the frequency domain.

This chapter describes two common ways of representing image data in the frequency domain. The first technique is via the discrete cosine transform (DCT), which is closely related to the second technique, the discrete Fourier transform (DFT). These two techniques are based on Fourier theory, which (in informal terms) states that any function can be expressed as the sum of a set of weighted sinusoids, which are known as basis functions. Since a pure sinusoid is fully characterized by its frequency (or wavelength), amplitude and offset (or phase), the task of representing an image in the frequency domain is to identify an appropriate set of wavelengths that could occur in the image and for each wavelength determine its corresponding amplitude. The DCT and DFT are techniques for taking a digital image that is represented in the spatial domain and representing that image in the frequency domain.

9.2 Image Frequency

Section 3.3 informally described image frequencies as the amount by which samples change over some spatial distance. Variations in light intensity over short

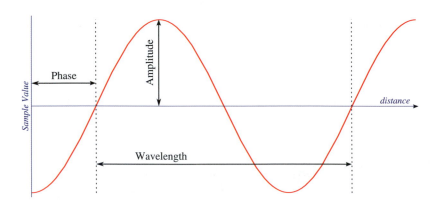

Figure 9.1. The properties of a sinusoid.

distances correspond to high frequencies while variations in light intensity over long distances correspond to low frequencies. Larger variations correspond to larger amplitude while smaller variations correspond to smaller amplitude. Image frequencies can be more formally described as complex sinusoidal patterns that occur within an image.

A sinusoidal function is shown in Figure 9.1, where the horizontal axis is a measure of spatial distance and the vertical axis corresponds to the luminance level at a point some distance from the leftmost point. The amplitude of the sinusoid corresponds to the maximum strength of the wave, the wavelength measures how quickly the light varies across distance, and the phase is simply an offset from the point of reference.

Equation (9.1) gives a useful mathematical definition of a real continuous sinusoidal function f. This function will serve as a basic building block in our understanding of the more formal definitions of image frequencies that occur later in this chapter. Equation (9.1) gives f as a function of spatial distance x that ranges over the interval 0 to N. The amplitude of the sinusoid is A, the phase is given by the real number ϕ and the frequency is given by real-valued u. In terms of image processing processing, function f can be understood as a continuous tone row profile of some scene that is projected onto a sensor of width N:

$$f(x) = A \cdot \sin(\frac{2\pi u x}{N} + \phi). \tag{9.1}$$

Variable u of Equation (9.1) controls the frequency of the sinusoid such that when $u = 1$ there is exactly one sinusoidal cycle as x ranges from 0 to n, when $u = 2$ there are two cycles, and so on. Variable u as described here is given in units of *cycles per image*, where the width of the image serves as the measure of unity length. Wavelength is inversely related to the frequency so that for a given u, the wavelength of the sinusoid is given as $1/u$ where image width as

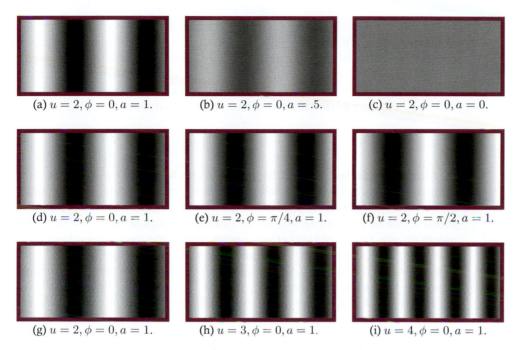

Figure 9.2. The effect of amplitude, phase, and frequency on sinusoidal images.

a unit measure. The offset, given in radians, serves to spatially shift the signal either left or right such that when the offset is zero the sinusoid is a sin wave and when the offset is $\pi/2$ the sinusoid is a cosine wave. The amplitude controls the strength or height of the sinusoid and is a dimensionless scalar.

Equation (9.1) is imaged in Figure 9.2 with varying values of frequency, amplitude, and phase. Note that while equation (9.1) is a one-dimensional function, the images in this figure are artificially extended into two dimensions for illustration only. The sinusoid has been scaled such that middle-gray corresponds to $f(x) = 0$, black corresponds to the case $f(x) = -A$, and white to the case $f(x) = A$.

In the top row of Figure 9.2 there are two cycles and hence $u = 2$; the phase is set to zero while the amplitude decreases from 1 in the first column to .5 in the middle and 0 in the third. Note that when the amplitude is set to zero the *sinusoid* becomes a scalar constant, exhibiting no change over distance. In the middle row the frequency and amplitude are fixed while the sinusoid is shifted by 0 radians in the first column, by $\pi/2$ radians in the middle, and $\pi/2$ in the third. The bottom row illustrates the effect of the frequency setting on the sinusoidal function while holding the phase and amplitude constant. The frequency setting ranges from two in the first column (note that there are two complete cycles) to three in the middle and four in the final column.

Consider sampling the continuous sinusoidal function of Equation (9.1) by taking N uniformly spaced samples such that the center of each sample is located at multiples of $\frac{1}{2}$ along the spatial dimension. This sampling is given in Equation (9.2), where x is now an integer in $[0, N-1]$ and corresponds with a column index in the row profile. Parameter u is a nonnegative scalar that controls the frequency of the sinusoid in the continuous domain:

$$f(x) = A \cdot \sin\left(\frac{2\pi u}{N}(x + 1/2) + \phi\right) \quad x \in [0, N-1]. \tag{9.2}$$

One of the fundamental effects of sampling is to place an upper limit on the frequencies that can be completely captured by sampled data. The Shannon-Nyquist sampling theorem establishes the limit by showing that the continuous sinusoid can be perfectly reconstructed if sinusoidal frequency is less than $\frac{1}{2}$ the sampling rate. Or stated another way, the sampling rate must be more than twice that of the sampled sinusoid. In digital image processing, the sampling rate is determined by the size of the image such that the highest frequency component of any $N \times N$ image is limited by $N/2$.

Figure 9.3 illustrates the reason for the Nyquist limit. A continuous sinusoid having $u = 1$ and $N = 8$ is uniformly sampled to generate the 8-element sequence shown in part (a). The samples are labeled such that x is in $[0, 7]$ and gives the value of the sinusoid, as denoted by the dots. Part (b) shows the results of sampling a higher frequency signal where $u = 7$ and $N = 8$. Note that the sampling rate of part (b) falls below the Shannon-Nyquist criteria, which implies that the sampling rate is insufficient to capture the relatively high frequency of the source signal. The inadequate sampling rate leads to a phenomena known as *aliasing*, which occurs when two separate signals become identical, or become aliases for each other, when sampled. Aliasing is illustrated in part (c) where the sinusoids of both (a) and (b) produce the same set of sampled data.

Equation (9.2) corresponds to a one-dimensional sampled sinusoid that has u cycles over a period of N samples where u is a nonnegative real value that is bounded by the Nyquist limit. A digital image is a two-dimensional structure that may have frequency components that vary not only horizontally but also

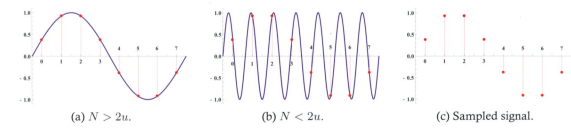

(a) $N > 2u$.   (b) $N < 2u$.   (c) Sampled signal.

Figure 9.3. Nyquist limit.

Figure 9.4. Invertibility.

vertically; hence Equation (9.2) must be extended in some fashion to accommodate both horizontal and vertical frequencies. Note that the horizontal and vertical frequencies are independent such that an image may have high frequency horizontal components and only low frequency vertical components.

In general terms we now claim that any $N \times N$ image can be represented as the weighted sum of sinusoidal images, or basis functions, corresponding to combinations of parameters u and v where u and v have been quantized to some discrete set of values. This claim is the central insight into the frequency domain representations presented in this text and it is essential to understanding the sections that follow. Given a set of basis functions, an image can be taken into the frequency domain by determining the weight of each basis functions.

While both the DCT and DFT are operations that convert spatial images into the frequency domain, each of these operations is invertible such that a spatial domain image can be reconstructed from its frequency domain representation. Figure 9.4 illustrates the concept of invertibility. In this figure, we see that an image in the spatial domain can be transformed into the frequency domain via the DCT. An image in the frequency domain can also be transformed into the spatial domain via the inverse DCT. It is important to note that there is precisely the same amount of *information* in each domain but the information is packaged in very different ways. In a theoretical sense, both the forward and inverse transforms fully preserve the information content and hence there is no information loss when converting between the spatial and frequency domains. When implementing these transforms computationally, however, information can be lost due to the finite precision of the computing system. Also note that while Figure 9.4 shows only the DCT, the DFT is also invertible.

Throughout this chapter we denote images given in the spatial domain by lower case symbols, and their corresponding frequency domain representations are given as cursive upper case symbols. Spatial image f, for example, is denoted in the frequency domain by $\mathcal{F}$. In addition it is sometimes convenient to describe the DCT and DFT as operators that take a spatial domain image as input and produce a frequency domain representation. Throughout this text we use the non-standard notation of an overlined calligraphic symbol to denote an operator that transforms an image from one domain into another. For example, $\bar{\mathcal{F}}$ is the

Notation	Meaning
f	spatial domain image
$\mathcal{F}$	frequency domain image
$\vec{\mathcal{F}}$	Fourier operator
$f(x, y)$	a spatial sample
$\mathcal{F}(u, v)$	a Fourier coefficient

Table 9.1. Notational conventions.

DFT operator that operates on spatial image f to produce the frequency domain representation $\mathcal{F}$ or, in other words, $\vec{\mathcal{F}}(f) = \mathcal{F}$.

Throughout this text we have also followed convention by denoting the column of a spatial image as x and the row of a spatial image as y. When accessing elements of a frequency domain image we denote the column by u and the row by v. Hence $f(x, y)$ represents the sample at location (x, y) in a spatial image while $\mathcal{F}(u, v)$ represents the amplitude of the sinusoid characterized by the parameters u and v. Table 9.1 gives a concise summary of the notation used throughout this chapter.

9.3   Discrete Cosine Transform

The discrete cosine transform (DCT) expresses an image as the sum of weighted cosines known as cosine basis functions. The wavelengths of these basis functions are determined by the dimensions of the source image while the amplitudes of the basis functions are determined by the samples themselves.

9.3.1   One-Dimensional DCT

In the one-dimensional case the DCT converts a sequence of N samples into a sequence of N DCT coefficients representing the amplitudes of the cosine basis functions. Letting $f(x)$ be a sequence of N samples the DCT coefficients are given by $\mathcal{C}(u)$ where $x, u \in [0, N-1]$. The N values $f(0), f(1), ..., f(N-1)$ are transformed into the N values $\mathcal{C}(0), \mathcal{C}(1), ..., \mathcal{C}(N-1)$ according to Equation (9.3):

$$\mathcal{C}(u) \quad = \quad \alpha(u) \sum_{x=0}^{N-1} f(x) \cos\left(\frac{(2x+1)u\pi}{2N}\right), \tag{9.3}$$

$$\alpha(k) \quad = \quad \begin{cases} \sqrt{\frac{1}{N}} & \text{if } k = 0, \\ \sqrt{\frac{2}{N}} & \text{otherwise.} \end{cases} \tag{9.4}$$

Notice that in Equation (9.3) each DCT coefficient is dependent on *every* source sample. Thus, if a single source sample is changed *every* DCT coefficient is altered. In other words, the information of a single spatial sample is distributed across all of the DCT coefficients. Parameter u determines the frequencies of the N cosine basis functions where larger values indicate higher frequency components and lower values correspond to lower frequency components. Parameter x serves to connect a source sample with a particular point on a basis function or, in more technically precise language, we say that f is projected onto the basis functions through x.

The lowest frequency component occurs whenever $u = 0$, signifying a wave of zero frequency (or infinite wavelength). Equation (9.3) then becomes

$$C(0) = \sqrt{\frac{1}{N}} \sum_{x}^{N-1} f(x). \tag{9.5}$$

Coefficient $C(0)$ is therefore a scaled average of all sample values and is commonly known as the *DC* coefficient while all others are known as the *AC* coefficients. These designations are derived from the field of signal processing where the DCT transform is used to analyze direct-current (DC) and alternating-current (AC) flow in electronic circuitry.

Consider a digital grayscale image with eight columns where a single row of the image contains the samples 20, 12, 18, 56, 83, 110, 104, 114. When the information from this single row is transformed into the frequency domain using the DCT technique, we obtain eight values that roughly correspond to the amplitudes of the eight harmonics given by the DCT formula. Since there are eight samples we know that $N = 8$ and u ranges from 0 to 7. In order to compute $C(0)$, for example, we apply Equation (9.3):

$$
\begin{aligned}
C(0) &= \sqrt{\tfrac{1}{8}} \sum_{x=0}^{7} f(x) \cos\left(\frac{(2x+1) \cdot 0\pi}{2 \cdot 8}\right) \\
&= \sqrt{\tfrac{1}{8}} \sum_{x=0}^{7} f(x) \cos(0) \\
&= \sqrt{\tfrac{1}{8}} \cdot \{f(0) + f(1) + f(2) + f(3) + f(4) + f(5) + f(6) + f(7)\} \\
&= .35 \cdot \{20 + 12 + 18 + 56 + 83 + 110 + 104 + 115\} \\
&= 183.14. \tag{9.6}
\end{aligned}
$$

Since parameter u extends over the interval $[0, 7]$ we must repeatedly apply Equation (9.3) for all values of u. Table 9.2 gives both the spatial representation and the corresponding DCT coefficients of the example row profile. Note that the DCT coefficients have a much larger range than the spatial samples and

f(0)	f(1)	f(2)	f(3)	f(4)	f(5)	f(6)	f(7)
20	12	18	56	83	110	104	114
$\mathcal{C}(0)$	$\mathcal{C}(1)$	$\mathcal{C}(2)$	$\mathcal{C}(3)$	$\mathcal{C}(4)$	$\mathcal{C}(5)$	$\mathcal{C}(6)$	$\mathcal{C}(7)$
183.1	-113.0	-4.2	22.1	10.6	-1.5	4.8	-8.7

Table 9.2. Numeric example of the DCT.

that some of the coefficients are negative. In addition, note that the first two coefficients are significantly larger in magnitude that the remaining six.

Implementation. Given a one-dimensional row profile of N samples, the DCT coefficients can be computed by direct encoding of Equation (9.3). Listing 9.1 gives a straightforward implementation for computing the DCT coefficients as an array of floats given a one-dimensional array of floats as input. The computational performance of the forwardDCT method is on the order of $O(N^2)$, assuming that the input array is of length N since the inner for loop executes N times for every one of the N iterations of the outer loop.

```
public static float[] forwardDCT(float[] data) {
  final float alpha0 = (float)Math.sqrt(1.0/data.length);
  final float alphaN = (float)Math.sqrt(2.0/data.length);
  float[] result = new float[data.length];

  for(int u=0; u<result.length; u++){
    for(int x=0; x<data.length; x++) {
      result[u] +=
        data[x]*(float)Math.cos((2*x+1)*u*Math.PI/(2*data.length));
    }
    result[u] = result[u] *( u == 0 ? alpha0 : alphaN) ;
  }
  return result;
}
```

Listing 9.1. One-dimensional DCT method.

9.3.2   Two-Dimensional DCT

Since digital images are two-dimensional data structures, the one-dimensional DCT must be extended to two dimensions in order to process digital images. Equation (9.7) gives the two-dimensional DCT transform for an $N \times N$ image where function α is the same as in Equation (9.3). Note that every DCT coefficient $\mathcal{C}(u, v)$ is dependent upon every sample of spatial image f. This implies that if a *single* sample in f is changed then every DCT coefficient is altered:

$$C(u,v) = \alpha(u)\alpha(v) \sum_{x=0}^{N-1} \sum_{y=0}^{N-1} f(x,y) \cos\left(\frac{(2x+1)u\pi}{2N}\right) \cos\left(\frac{(2y+1)v\pi}{2N}\right).$$

$$(9.7)$$

Also note that the cosine terms are dependent only on parameters u and v where we assume that N is a constant in this context. Each (u,v) pair then establishes a basis, or kernel, and the corresponding DCT coefficient is a measure of how dominant that basis is in the spatial image. Since there are N^2 combinations of u and v there are N^2 bases or kernels that contribute to the spatial image under analysis. Again, the central notion of the DCT transform is that any spatial image can be constructed by scaling and summing these N^2 base images.

Figure 9.5 illustrates the DCT basis functions and corresponding kernels for a 4×4 image. The cosine basis functions are plotted in (a) and are positioned within the table at (u,v). Note that the functions in row zero, and column zero are identical except for rotation since one is aligned horizontally with the source image and the other is aligned vertically with the source image. The kernels given in (b) are sampled and quantized images of the basis functions where black corresponds to negative 1, mid-gray corresponds to zero and white corresponds to positive 1.

One important property of the DCT is that it is *separable*. Separability here means that the DCT coefficients of the row and column profiles are independent, or separable, and that the 2D coefficients can be computed by a two-pass

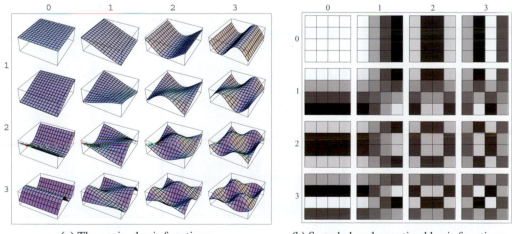

(a) The cosine basis functions.   (b) Sampled and quantized basis functions.

Figure 9.5. DCT basis functions.

application of the one-dimensional DCT transform. The DCT is also invertible and hence a spatial domain image can be exactly recovered from the DCT coefficients. The inverse discrete cosine transform for both the one-dimensional and two-dimensional cases are given as

$$f(x) = \sum_{u=0}^{N-1} \alpha(u)\mathcal{C}(u) \cos\left(\frac{(2x+1)u\pi}{2N}\right), \tag{9.8}$$

$$f(x,y) = \sum_{u=0}^{N-1}\sum_{v}^{N-1} \alpha(u)\alpha(v)\mathcal{C}(u,v) \cos\left(\frac{(2x+1)u\pi}{2N}\right)$$
$$\times \cos\left(\frac{(2y+1)v\pi}{2N}\right). \tag{9.9}$$

The discrete cosine transform is said to provide a high degree of energy compaction. Much like a garbage compactor takes a container of trash and compresses it into a much smaller volume by eliminating empty space, the DCT transform takes the spatial domain samples and compresses the information into a more compact form. The central idea in energy compaction is more specifically that the lower frequency DCT coefficients tend to be relatively large while the higher frequency coefficients tend to be relatively small. In other words, most of the information of the spatial image is compacted into the lower frequency terms of the DCT. In the two dimensional case this implies that those coefficients in the upper-left will typically be orders of magnitude larger than the coefficients in the lower and right regions.

Certain compression techniques leverage the energy compaction property of the DCT by converting an image into its DCT coefficients and then discarding

(a) Source image.   (b) Recovered with 1%.   (c) Recovered with 5%.

Figure 9.6. Use of DCT coefficients for compression.

the relatively insignificant higher frequency coefficients. JPEG compression is based on the DCT and is discussed further in Chapter 10. Figure 9.6 illustrates the energy compaction property of the DCT. In this figure a source image (a) is converted into DCT coefficients. The image is then converted back into the spatial domain using only 1% of the DCT coefficients (b) and 5% of the DCT coefficients (c). The degree of perceptual information that is retained when discarding 95% of the DCT coefficients is rather astonishing. When discarding information in the DCT domain, the high-frequency components are eliminated since they contain less perceptually relevant information than the low-frequency components. This results in an image that is blurred and also creates artificial ringing along edges as can clearly be seen in (b).

Implementation. Given an $N \times N$ image, the DCT coefficients can be computed in time proportional to $O(N^4)$ using a straightforward brute-force implementation of Equation (9.7). The separability property of the DCT, however, allows the computation to be easily implemented in terms of the one-dimensional transform and also results in a runtime proportional to $O(N^3)$. Listing 9.2 gives a

```
1  public static float [][] forwardDCT(float [][] data) {
2      float [][] result = new float[data.length][data.length];
3
4      // 1D DCT of every row
5      for(int u=0; u<result.length; u++) {
6          result[u] = forwardDCT(data[u]);
7      }
8
9      // 1D DCT of every column of intermediate data
10     float [] column = new float[data.length];
11     for(int v=0; v<result.length; v++) {
12         for(int row=0; row<data.length; row++){
13             column[row] = result[row][v];
14         }
15
16         float [] temp = fowardDCT(column);
17         for(int row=0; row<data.length; row++){
18             result[row][v] = temp[row];
19         }
20     }
21
22     return result;
23  }
```

Listing 9.2. Brute force 1D DCT.

complete implementation of the two-dimensional forward DCT. This method first applies the one-dimensional DCT to every row of the input samples and then takes the intermediate result and applies the one dimensional DCT to every column. While this listing assumes that the input data is square in dimension it can be easily altered to accurately compute the DCT coefficients for data of any width and height.

Note that the cosine term is dependent only on variables u and x, each of which is constrained to the interval $[0, N - 1]$. Significant computational effort can be saved by pre-computing these N^2 cosine terms and reusing them when performing the row and column transformations.

9.4   Discrete Fourier Transform

The discrete Fourier transform (DFT) expresses an image as the sum of weighted sinusoids known as the sinusoidal basis functions. As with the DCT, the wavelengths of these basis functions are determined by the dimensions of the source image while the amplitudes of the basis functions are determined by the samples themselves. The DFT is a generalization of the DCT and differs from the DCT in that the basis functions themselves lie in the complex plane and generate complex coefficients rather than the real-valued coefficients of the DCT.

In the one-dimensional case the DFT converts a sequence of N samples into a sequence of N DFT complex coefficients where the magnitude of the complex coefficient represents the amplitudes of the corresponding basis functions. Letting $f(x)$ be a sequence of N samples, the DFT coefficients are given by $\mathcal{F}(u)$ where $x, u \in [0, N - 1]$. The N values $f(0), f(1), ..., f(N - 1)$ are transformed into the N coefficients $\mathcal{F}(0), \mathcal{F}(1), ..., \mathcal{F}(N - 1)$ according to

$$\mathcal{F}(u) = \frac{1}{N} \sum_{x=0}^{N-1} f(x) \left[\cos\left(\frac{2\pi u x}{N}\right) - j \sin\left(\frac{2\pi u x}{N}\right) \right], \qquad (9.10)$$

where the symbol j is used to denote the *imaginary unit* and satisfies $j^2 = 1$.

Again notice that in equation (9.10) each DFT coefficient is dependent on every source sample. Thus, if a single source sample is changed every DFT coefficient is altered. In other words, the information of a single spatial sample is distributed across all of the DFT coefficients. The variable u determines the frequency of the N basis functions where larger values indicate higher frequency components and lower values correspond to lower frequency components. The variable x serves to connect a source sample with a particular point on a basis function or, in more technically precise language, we say that f is projected onto the basis functions.

Equation (9.10) is often written in more compact form by applying Euler's formula,[1] which states that for any real number x,

$$\cos(x) - j\sin(x) = e^{-jx},$$
(9.11)

where e is defined to be the base of the natural logarithm, a real-valued number approximately equal to 2.71828. Equation (9.10) can thus be rewritten as

$$\mathcal{F}(u) = \frac{1}{N} \sum_{x=0}^{N-1} f(x) e^{-j2\pi ux/N}.$$
(9.12)

Since the DFT is an invertible transform, the original sequence of coefficients can be exactly recovered from the DFT coefficients. The one-dimensional inverse transform has nearly the same form as the forward transform and is given as

$$f(x) = \frac{1}{N} \sum_{u=0}^{N-1} \mathcal{F}(u) e^{j2\pi ux/N},$$
(9.13)

where only the sign of the exponent differs from the forward transform. While the information contained by the N samples of f is identical to the information given by the N DFT coefficients of $\mathcal{F}$ the information is simply represented in different forms. Additionally, while information may be lost when using computing machinery to convert between the spatial and frequency domains, this loss is an artifact of machine precision and not intrinsic to the mathematical formulation itself. This one dimensional transform can be extended into two dimensions. The two-dimensional forward and inverse DFT transforms of an $N \times N$ digital image are given by

$$\mathcal{F}(u, v) = \frac{1}{N} \sum_{x=0}^{N-1} \sum_{y=0}^{N-1} f(x, y) e^{-j2\pi(ux+vy)/N},$$
(9.14)

$$f(x, y) = \frac{1}{N} \sum_{u=0}^{N-1} \sum_{v=0}^{N-1} \mathcal{F}(u, v) e^{j2\pi(ux+vy)/N}.$$
(9.15)

9.4.1　Image Spectra

Each DFT coefficient $\mathcal{F}(u, v)$ is a complex value lying in the complex plane and hence when converting a two-dimensional raster of samples into DFT coefficients we end up with a two dimensional array of complex values. A complex

---

[1]Euler's formula is named after Leonhard Euler, who discovered this relationship in the mid 1700s.

value can be expressed in either polar or Cartesian coordinates. When expressed in Cartesian coordinates the DFT coefficients have the form

$$\mathcal{F}(u,v) = R(u,v) + jI(u,v),$$

where $R(u,v)$ is the real and $I(u,v)$ is the imaginary part of the complex number. The DFT coefficients are, however, more commonly and usefully given in polar coordinates and have the form

$$\mathcal{F}(u,v) = |F(u,v)|e^{j\phi(u,v)},$$

where $|\mathcal{F}(u,v)|$ is known as the magnitude and $\phi(u,v)$ is known as the phase. The magnitude of a complex number is simply a measure of how far the point lies from the origin in the complex plane while the phase is an angular measure indicating in which direction the point lies in the complex plane. The phase thus falls in the range $[-\pi, +\pi]$ and is cyclic since $-\pi$ is equivalent to $+\pi$ in angular terms. The magnitude and phase can be obtained from the real and imaginary parts of a complex number such that

$$\begin{aligned} |\mathcal{F}(u,v)| &= \sqrt{R^2(u,v) + I^2(u,v)}, \\ \phi(u,v) &= \tan^{-1}\left[\frac{I(u,v)}{R(u,v)}\right]. \end{aligned}$$

The complex DFT coefficients can then be split into a raster of magnitudes and a raster of phases. These rasters can be rendered as images as illustrated in Figure 9.7. In this figure a source image has been transformed into DFT coefficients and then split into its corresponding magnitudes and phases, which have been rendered as the grayscale images shown in (b) and (c).

 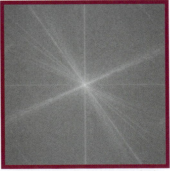 

(a) Source image f.   (b) Amplitude spectrum, $|\mathcal{F}|$.   (c) Phase spectrum, ϕ.

Figure 9.7. DFT Spectrum.

It should be noted that since the dynamic range of the magnitude values is so large, it is common to log compress the magnitudes prior to rendering. The scale of image (b) is therefore logarithmic where bright samples indicate a relatively large magnitude and dark areas indicate a correspondingly small magnitude. The phase data is more difficult to interpret since the data is cyclic and there appears to be no discernible pattern. Although we may be tempted to believe that the phase data is therefore of little importance this is not correct. The phase data contains vital information regarding the spatial location of the sinusoidal basis functions.

In addition, the images of Figure 9.7 have been shifted so that the DC element is centered, which implies that low frequency elements are situated nearer to the center while the frequency increases with distance from the center of the image. While this arrangement differs from what is presented in Equations (9.14), where the DC term located at location $(0, 0)$, it is the conventional method for visualizing the amplitude and phase spectra and all spectral images throughout this text adopt this convention.

The magnitudes are commonly known as the spectrum of an image but this is a rather loose definition. More precisely we say that the magnitudes constitute the amplitude spectrum and the phases constitute the phase spectrum. The power spectrum is another common term that is defined as the square of the amplitude spectrum. The power spectrum $P(u, v)$ is given by

$$P(u, v) = |\mathcal{F}(u, v)|^2 = R^2(u, v) + I^2(u, v).$$

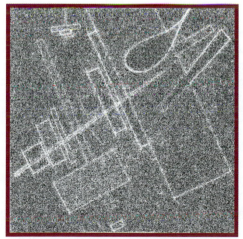
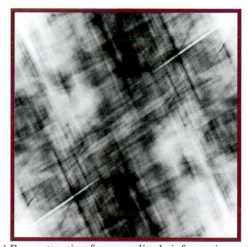

(a) Reconstructed from phase information only.   (b) Reconstruction from amplitude information only.

Figure 9.8. Comparison of the contribution of the amplitude and phase spectrum.

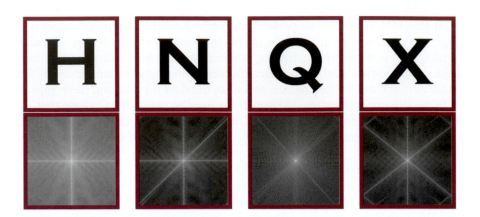

Figure 9.9. Illustration of DFT properties.

Both the magnitude and phase spectra contain vital information regarding the source image. Consider Figure 9.8, where the source image has been re-constructed from the DFT coefficients after destroying the amplitude spectrum while keeping the phase spectrum intact (a), or after destroying the phase spectrum while keeping the amplitude spectrum intact (b). The amplitude spectrum was destroyed by setting each magnitude to unity while the phase spectra was destroyed by setting each phase to zero.

While it is difficult to gain much insight from a visual examination of the amplitude spectrum, there are some points to consider. A strong edge in the source image will generate a strong edge in the spectrum as well, although rotated $90°$ from the source. Strong elliptical edges in the source image will also carry over as strong elliptical edges in the spectrum. Figure 9.9 shows four source images and their corresponding spectra. Note that the letter H generates strong vertical and horizontal edges in the spectrum since the letter H itself has strong vertical and horizontal edges. Also note that the letter Q generates elliptical edges in the spectrum since the letter Q is itself of circular shape.

The magnitude spectrum contains information about the shape of objects in the source while the phase spectrum encodes the number of objects and where they are located in the source. Consider, for example, a source image that has many randomly placed copies of the letter Q. The magnitude spectrum would be the same as if there were a single letter Q somewhere in the source while the phase spectrum would be changed to indicate where the letters actually show up in the source. It is also important to note that the center of the magnitude plot corresponds to the average (or DC) component while the sinusoidal basis functions increase in frequency with increasing distance from the center.

9.4.2 Properties of the DFT

The DFT of an image possesses important properties that can be exploited when performing image processing tasks. In this section we consider the way in which the DFT of an image will change if we alter the source image in specific ways.

Translation, Rotation, and Distributivity. Translation of the source image will cause the phase spectrum to change but will leave the magnitude spectrum unchanged. This can be understood at an intuitive level by recognizing that the phase spectrum contains information about the location of objects in the source while the magnitude spectrum contains information about the object's shape. Under translation the location of an object is altered while the shape is unchanged and hence the magnitude spectrum is unchanged but the spectra is linearly shifted.

Rotation of a source image f by an angle θ results in a rotation of the spectra by an amount θ. Since the sinusoidal basis functions are used to reconstruct the shapes in the source image, they must be rotated in order to reconstruct the rotated shape.

The Fourier transform and its inverse are distributed over addition but not multiplication. This means that if two source images are summed in the spatial domain the corresponding spectrum is the sum of the individual source spectra. This property is given mathematically as

$$\vec{\mathcal{F}}(f + g) = \vec{\mathcal{F}}(f) + \vec{\mathcal{F}}(g),$$

where recall that the notation $\vec{\mathcal{F}}$ is used to denote an operator that accepts a spatial domain image f and converts it into the frequency domain using the discrete Fourier transform.

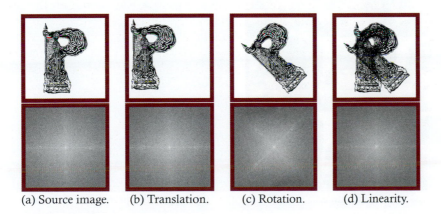

(a) Source image.   (b) Translation.   (c) Rotation.   (d) Linearity.

Figure 9.10. Properties of the DFT under translation, rotation, and linear combination.

Figure 9.10 illustrates these properties. In this figure the magnitude spectra corresponding to four source images are shown. Image (a) is the reference image. This reference image has been translated in (b) and rotated in (c). The original reference and the rotated reference are then added together to obtain the image of (d). Note that the amplitude spectra of (a) and (b) are identical while the amplitude spectrum of (c) is obtained by rotating the spectrum of (a). The amplitude spectrum of (d) is obtained by adding the amplitude spectra of (b) and (c).

Periodicity. The Fourier transform views a spatial source image as an infinitely tiled image covering all space in a fashion identical to the circular indexing described in 6.2.1. This follows from the idea that the sinusoidal basis functions are themselves periodic and hence everywhere defined. For a $M \times N$ source image we can state this periodicity property very concisely as

$$\mathcal{F}(u,v) = \mathcal{F}(u+M,v) = \mathcal{F}(u,v+N) = \mathcal{F}(u+M,v+N), \qquad (9.16)$$

While this property of the DFT may initially seem to be of little consequence, it carries important implications. One of the most significant is that since the source image wraps around itself, the source image will appear to have edges where no edges actually exist. If, for example, the left side of an image is bright and the right side is dark, there will be a false edge created when the image is tiled since the brightness of the left side is placed adjacent to the darkness of the right side. These artificial edges will then be reflected in the amplitude spectrum and give a false characterization of the power spectrum of the source image.

(a) Tiled source.

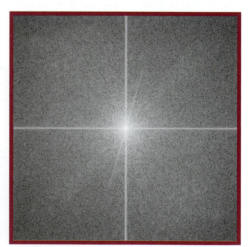

(b) Amplitude spectrum of the source.

Figure 9.11. Periodicity effects of the DFT.

Figure 9.11 illustrates how the DFT can present a false view of the spectrum due to the fact that the source image is tiled. In this example, a source image is implicitly assumed to be tiled over all space which causes false edges to appear at the left-right seams as well as at the top-bottom seams in the tiling. These false edges are reflected in the strong vertical and horizontal lines in the amplitude spectrum.

9.4.3   Windowing

Windowing is a technique to minimize the artificial distortions that are injected into the spectrum of an image due to tiling discontinuities. Windowing is an image processing filter that is applied prior to conversion into the Fourier domain and an inverse windowing filter must be applied when recovery of the source image is desired from the DFT coefficients.

Since the spectrum is distorted by discontinuities at the edge of an image when the image is tiled, windowing forces the edge samples to be nearly identical across seams. This is accomplished by multiplying each sample of the source image by a scaling factor that is unity near the center of an image and falls smoothly to zero near the borders. As a result, all border pixels are nearly black and artificial discontinuities are largely eliminated. The inverse windowing procedure is to simply divide each sample by the same scaling factors hence restoring the original values except for those few samples that have been truncated to zero by the windowing process.

Numerous windowing functions are in common use; we will define four of the more common windowing functions here. The Bartlett windowing function is the simplest, representing a linear decay from unity at the center of the source image to zero at the edges of an image. The Bartlett windowing function is given as

$$w(r) = \begin{cases} 1 - \frac{r}{r_{\max}} & r \leq r_{\max}, \\ 0 & r > r_{\max}, \end{cases} \tag{9.17}$$

where r is the Euclidean distance from the center of the image and $r_{\max}$ is the minimum of the height and width of the source image. The Hanning window is smoother than the Bartlett and hence has superior mathematical properties. All of the windowing functions are zero-valued for $r > r_{\max}$ but differ in their manner of decay as they move toward zero. The Hanning window, for $r \leq r_{\max}$ is given as

$$w(r) = \frac{1}{2} - \frac{1}{2} \cos \left[\pi (1 - \frac{r}{r_{\max}}) \right]. \tag{9.18}$$

The Blackman window is more complex and somewhat narrower than the Hanning window and is defined as

$$w(r) = .42 - \frac{1}{2} \cos \left[\pi \left(1 - \frac{r}{r_{\max}} \right) \right] + .08 \cos \left[2\pi \left(1 - \frac{r}{r_{\max}} \right) \right]. \tag{9.19}$$

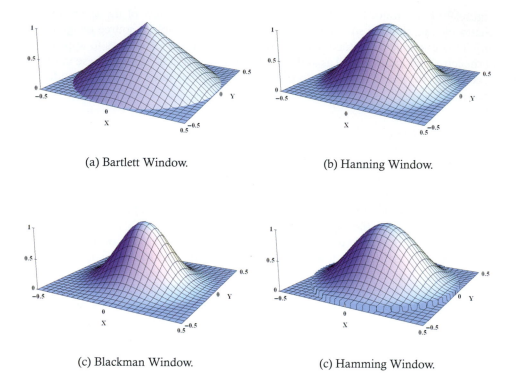

(a) Bartlett Window.   (b) Hanning Window.

(c) Blackman Window.   (c) Hamming Window.

Figure 9.12. Normalized windowing functions covering a 1×1 region.

While the Hamming window is smooth, it does not decay to zero at the edges and is given as

$$w(r) = 0.53836 + 0.46164 \cos\left(\pi \frac{r}{r_{\max}}\right). \tag{9.20}$$

These four window functions are plotted in Figure 9.12. In these plots, the values of x and y are normalized over the interval $[-1/2, 1/2]$ and, of course, the values of the windowing function range from 0 to 1.

Figure 9.13 illustrates the effect of windowing on the spectrum. In this example, a source image is windowed such that the edges of the image are all nearly black so that while the source image is implicitly assumed to be tiled over all space, there are no edges present at the seams. Windowing the source prior to taking the DFT coefficients strongly dampens any edge effect as can be seen in the magnitude spectrum, especially when compared to the magnitude spectrum of Figure 9.11.

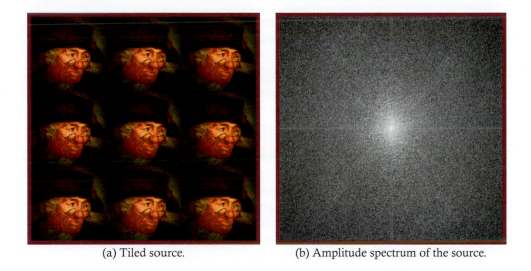

(a) Tiled source. (b) Amplitude spectrum of the source.

Figure 9.13. The effect of windowing on the spectrum.

9.4.4 Frequency Domain Filtering

We have previously shown that images can be filtered in the spatial domain by directly examining and changing sample values. Of course image filtering can be accomplished in the frequency domain by directly examining and altering the DFT coefficients as well. Frequency domain filtering often has advantages over spatial domain filtering in terms of both computational efficiency and utility. Frequency domain filtering can be generalized as the multiplication of the spectrum $\mathcal{F}$ of an image by a transfer function $\mathcal{H}$. In other words

$$\mathcal{G}(u,v) = \mathcal{F}(u,v) \cdot \mathcal{H}(u,v), \tag{9.21}$$

where $\mathcal{F}$ is the DFT of a source image f, $\mathcal{H}$ is the transfer function (or filtering function), and $\mathcal{G}$ is the DFT that results from application of the filter. All of these entities, $\mathcal{F}$, $\mathcal{H}$, and $\mathcal{G}$ are complex-valued, and hence the multiplication of $\mathcal{F}$ with $\mathcal{H}$ is the multiplication of complex values which may change both the magnitude and phase of the source DFT. In practice, however, most frequency domain filters are zero-phase-shift filters that simply means that they leave the phase spectrum unaltered while modifying the amplitude spectrum.

Convolution, as we have seen, is one of the most powerful filtering techniques in image processing and can be used to either blur or sharpen an image. The convolution theorem states that convolution can be performed in the frequency domain by straightforward multiplication of two image transform. Given spatial image f and kernel h and having corresponding DFTs of $\mathcal{F}$ and $\mathcal{H}$, the convolution theorem states that

$$f \otimes h \Leftrightarrow \mathcal{F} \cdot \mathcal{H}. \tag{9.22}$$

In other words, convolution in the spatial domain is equivalent to point-by-point multiplication of the DFT coefficients.

This is an important theorem with strong implications for implementing an image convolution filter. Recall that given an $M \times M$ kernel and a $N \times N$ image, the computational effort is on the order of a whopping $O(M^2 N^2)$. The convolution theorem states that this operation can be accomplished in the frequency domain in time proportional to $O(N^2)$ since there are only N^2 complex multiplications to perform. To achieve this level of performance we must go through a number of processes that will take an image into the frequency domain, directly modify the magnitude spectrum, and then take the image back into the spatial domain. In other words, to convolve a spatial domain image f with kernel h we must perform the following four steps:

1. Compute the DFT coefficients of f to obtain $\mathcal{F}$.

2. Compute the DFT coefficients of h to obtain $\mathcal{H}$.

3. Compute the product of these two transforms to obtain $\mathcal{F} \cdot \mathcal{H}$.

4. Compute the inverse DFT of the product to obtain $\vec{\mathcal{F}}^{-1}(\mathcal{F} \cdot \mathcal{H})$.

At first glance this four-step process seems to require more computational effort than a straightforward brute-force spatial domain convolution. As we will see, however, if the DFT can be computed efficiently enough, this process yields far superior runtime performance than the spatial domain implementation. It should also be noted that the frequency domain kernel $\mathcal{H}$ can often be generated directly in the Fourier domain without resorting to the DFT to convert its spatial domain representation into the frequency domain.

Low pass filtering. Low pass filtering attenuates high frequency components of an image while leaving low frequency components intact. Low pass filtering can be very easily and naturally accomplished in the frequency domain since the frequency components are explicitly isolated in the spectrum. Recall that the DFT coefficients correspond to frequency components of the source and that the frequency increases with increasing distance from the center of the shifted spectrum. Low pass filtering is then accomplished by zeroing the amplitude of the high frequency DFT coefficients. Determining which DFT coefficients should be zeroed amounts to determining how far the coefficient is from the center of the shifted spectrum. Typically, a threshold radius is chosen such that all DFT coefficients outside of this threshold radius have their magnitude set to zero while all DFT coefficients falling inside of the threshold are unchanged. If the cutoff radius is large then fewer DFT coefficients are erased and hence less blurring occurs. Of course, if the cutoff radius is small then a high proportion of the DFT coefficients are erased and the blurring effect is more pronounced.

We then seek a transfer function that will correspond to this low pass filtering technique. The ideal low pass filter represents perhaps the simplest of these low pass filters. The ideal low pass transfer function is given as

$$\mathcal{H}(u, v) = \begin{cases} 1 & \sqrt{u^2 + v^2} \leq r_c, \\ 0 & \text{otherwise,} \end{cases} \tag{9.23}$$

where r_c is known as the cutoff frequency, which determines the strength of the filter.

This is illustrated in Figure 9.14. A source image has been converted into the frequency domain after which the DFT coefficients have undergone point-by-point multiplication with an ideal filter. The result is that high frequency portions of the magnitude spectrum have been truncated. The spatial-domain image that is recovered after truncating the DFT coefficients is shown as the blurry image of (b).

The term *ideal* as it is used of an ideal low pass filter should not be taken to mean that this filter is optimal or the most desirable for low pass filtering. An ideal low pass filter is ideal or perfect in the sense that it has an exact cutoff above which all terms are exactly zero and below which all terms are set to unity. An ideal low pass filter, far from being desirable for image processing, is rarely used

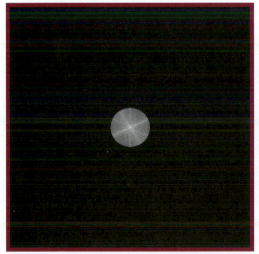 

(a) Filtered magnitude spectrum.   (b) Reconstructed spatial domain image.

Figure 9.14. Ideal low-pass filtering in the frequency domain.

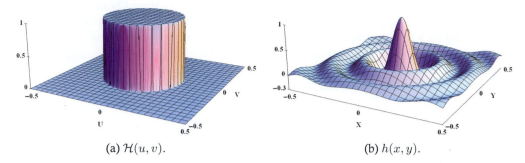

(a) $\mathcal{H}(u, v)$.   (b) $h(x, y)$.

Figure 9.15. Ideal low-pass filter in (a) the frequency domain and (b) the spatial domain.

in practice due to a very undesirable side effect known as *ringing* as can be seen in Figure 9.14. The blurred image that results from convolution with an ideal low pass filter is also distorted by an obvious rippling at object edges.

The source of blurring can be seen by considering the transfer function in both the frequency and spatial domains. In the Fourier domain the ideal low pass filter can be visualized as a cylindrical surface as illustrated in part (a) of Figure 9.15. This function can be transformed, via the inverse DFT, into the spatial domain to obtain the function shown in part (b). The spatial domain representation of the transfer function is known as the *sinc* function and is given as $sinc(r) = \sin(r)/r$ where r is distance from the origin. In order to convolve an image with the ideal transfer function in the spatial domain, the sinc function must be sampled, and quantized to obtain a kernel. As is evident from the figure, the kernel will have large values near the center and concentric circles of alternating negative and positive values near the extremities of the kernel. The positive coefficients at the center provide the desired blurring effect while the concentric circles near the periphery generate a rippling effect when passed over image edges.

While the ideal low pass filter is simple to implement, it is undesirable due to ringing at the edges. Since the ringing is caused by the sharp discontinuity at the cylinder edge, more practical transfer functions have been discovered that retain the general shape of the ideal low pass filter but are everywhere continuous and hence significantly reduce ringing at the boundaries. One of the most popular low pass filters is the Butterworth filter, which is a continuous function falling gradually to zero near the cutoff frequency. The Butterworth transfer function is given as

$$\mathcal{H}(u, v) = \frac{1}{1 + [r(u, v)/r_c]^{2n}}, \tag{9.24}$$

where once again r_c is the cutoff frequency and $r(u, v)$ is the Euclidean distance between the origin of the shifted spectrum and the point (u, v) in frequency

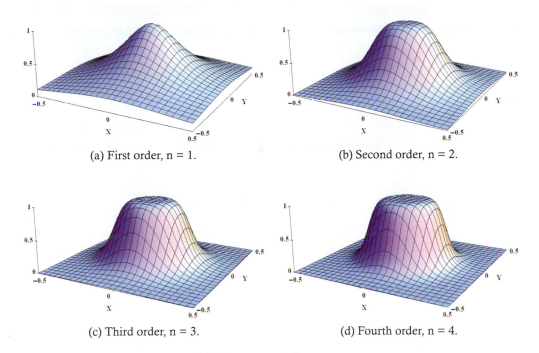

(a) First order, n = 1.
(b) Second order, n = 2.

(c) Third order, n = 3.
(d) Fourth order, n = 4.

Figure 9.16. Butterworth filters.

space, that is, $r(u, v) = \sqrt{u^2 + v^2}$. The Butterworth transfer function is parameterized on n, which occurs in the exponent and defines the order the filter. This transfer function hence represents a family of related transfer functions where lower order Butterworth filters are smoother and fall more gradually to zero at the cylinder boundary while higher-order Butterworth filters have sharper edges, falling quickly to zero at the cylinder boundaries.

Figure 9.16 shows Butterworth filters of orders 1 through 4 as they appear in the frequency domain over a normalized interval in u and v of $[-1/2, 1/2]$. Note that the cutoff radius is not the point at which the functions decay to zero, as was the case with the ideal low pass filter, but it is the point at which the transfer function drops to $1/2$. It is interesting to note that as the value of n approaches infinity, the Butterworth filter approaches the ideal low pass filter.

Figure 9.17 highlights the contrast between the ringing of the ideal low pass filter and the use of a Butterworth low pass filter. An ideal low pass filter having a cutoff located at a normalized 15% is compared to a fifth order Butterworth filter having the same cutoff frequency. While some small ringing artifacts are discernible in (b) the ringing has clearly been reduced from that of the ideal low pass filter in (a). In each image the primary objective of the filter, blurring, is comparable.

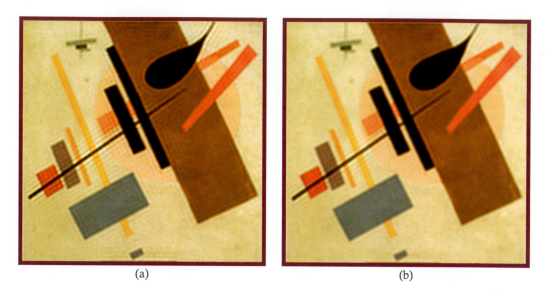

(a)   (b)

Figure 9.17. (a) Low pass filtering in the frequency domain using an ideal filter and (b) 5th order Butterworth filter. The cutoff radius was set at 15% in each example.

Other well-known frequency domain low pass filters include the Chebyshev and the Gaussian transfer functions. The Gaussian low pass filter has the very interesting property of having the same form in both the Fourier and spatial domains. In other words, the DFT of a Gaussian function is itself a Gaussian function. A Gaussian low pass filter introduces no ringing when applied either in the spatial or frequency domains. The Gaussian transfer function is given as

$$\mathcal{H}(u,v) = e^{-\frac{1}{2}[r(u,v)/r_c]^2}.\tag{9.25}$$

High pass filtering. High pass filtering attenuates low frequency components of an image while leaving high frequency components intact. High pass filtering, like low pass filtering, can be very naturally accomplished by direct manipulation of the DFT coefficients. The technique for performing high pass filtering is identical; however, the transfer functions are simply inverses to the low pass functions. The ideal high pass function is given by

$$\mathcal{H}(u,v) = \begin{cases} 0 & \sqrt{u^2+v^2} \leq r_c, \\ 1 & \text{otherwise.} \end{cases}\tag{9.26}$$

The ideal high pass suffers from the same ringing artifact as the ideal low pass filter and hence other smoother filters are generally used. The Butterworth high pass filter is given by

$$\mathcal{H}(u,v) = \frac{1}{1 + [r_c/r(u,v)]^{2n}},$$ (9.27)

and the Gaussian high pass filter by

$$\mathcal{H}(u,v) = 1 - e^{-\frac{1}{2}[r(u,v)/r_c]^2}.$$ (9.28)

Band filters. Low pass filtering is useful for reducing noise but may produce an image that is overly blurry. High pass filtering is useful for sharpening edges but also accentuates image noise. Band filtering seeks to retain the benefits of these techniques while reducing their undesirable properties. Band filtering isolates the mid-range frequencies from both the low-range and high-range frequencies. A band stop (or notch) filter attenuates mid-level frequencies while leaving the high and low frequencies unaltered. A band pass filter is the inverse of a band stop; leaving the mid-level frequencies unaltered while attenuating the low and high frequencies in the image. A band of frequencies may be conveniently specified by giving the center frequency and the width of the band. The band width determines the range of frequencies that are included in the band.

A band stop filter is essentially a combination of a low and high pass filter, which implies that ideal, Butterworth, and Gaussian band stop filters can be defined. Figure 9.18 illustrates the Butterworth band filters given as

$$\mathcal{H}(u,v) = \frac{1}{1 + [\Omega \cdot r(u,v)/(r(u,v)^2 - r_c^2)]^{2n}},$$ (9.29)

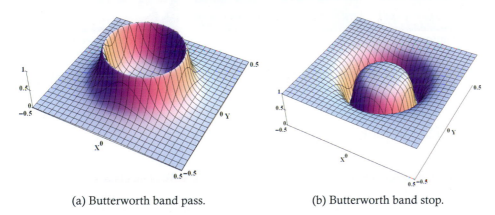

(a) Butterworth band pass.   (b) Butterworth band stop.

Figure 9.18. Normalized Butterworth second-order band filters having a center frequency of .4 and a band width of .3.

where Ω is the width of the band and r_c is the center of the band. The corresponding Butterworth band pass filter is given as

$$\mathcal{H}(u, v) = 1 - \frac{1}{1 + [\Omega \cdot r(u, v) / (r(u, v)^2 - r_c^2)]^{2n}}.$$ (9.30)

Periodic noise removal. Direct frequency domain filtering can be used to identify and eliminate periodic noise from a digital image. Periodic noise can penetrate a digital image if the acquisition system is in close proximity to electromechanical machinery, electric motors, power lines, and the like. Periodic noise manifests as sharp localized spikes in the spectrum, and hence specialized filters or interactive systems can be used to identify and eliminate this type of noise. Figure 9.19 shows how periodic noise can be identified and eliminated. A source image has been corrupted by some electromechanical noise source, as shown in (a). The corresponding spectrum shows a number of intense spikes that can be easily identified and truncated via direct manipulation of the spectrum. After this filtering process, the recovered image is shown in (c).

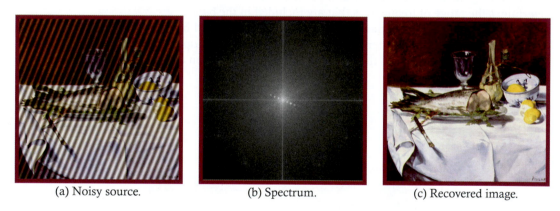

(a) Noisy source.   (b) Spectrum.   (c) Recovered image.

Figure 9.19. Periodic noise removal.

9.4.5   Implementation

Since working in the frequency domain requires manipulation of complex quantities, a Complex class is developed (see Listing 9.3). The class supports operations such as addition, subtraction, multiplication, and division. The class also enables clients to view the phase and magnitude of the quantity without any change in the underlying representation. This class is also *immutable* since there are no mutator methods; once a Complex object is created its internal state is

```
1   public class Complex {
2     protected float imaginary, real;
3
4     public Complex() {
5       imaginary = real = 0f;
6     }
7
8     public Complex(float re, float im) {
9       imaginary = im;
10      real = re;
11    }
12
13    public Complex plus(Complex c) {
14      return new Complex(c.getReal() + getReal(),
15                         c.getImagniary() + getImaginary());
16    }
17
18    public float getMagnitude() {
19      return (float)Math.sqrt(getReal() * getReal() +
20                              getImagnary() * getImaginary());
21    }
22
23    public float getPhase() {
24      return (float)Math.atan2(getImaginary(), getReal());
25    }
26
27    // other methods not shown
28  }
```

Listing 9.3. `Complex`.

unchangeable. While this design decision creates a slight degradation in performance, we choose this approach to increase reliability, specifically the likelihood of developing correct code.

Since the DFT algorithm requires complex-valued samples, we must author methods to convert a `BufferedImage` into a set of `Complex` values. We will also need a corresponding algorithm to convert the set of `Complex` values into a `BufferedImage` for visualization. Listing 9.4 implements a method that converts an image into a two-dimensional array of `Complex` objects. These complex values are still represented in the spatial domain but can be taken into the frequency domain via application of the forward DFT and manipulated directly in the frequency domain.

Given the `Complex` class we can now author a method to perform both the forward and inverse discrete Fourier transform. Listing 9.5 is a complete method that accepts a two dimensional array of complex values (i.e., an image) and applies the DFT. The type of DFT performed is indicated by the enumerated type

```
1  private Complex[][] imageToComplex(BufferedImage image) {
2     if(image == null) return null;
3     writableRaster input = image.getRaster();
4     Complex[][] data = new Complex[image.getHeight()][image.getWidth()];
5     for (int y =0; y < image.getHeight(); y++) {
6        for (int x =0; x < image.getWidth(); x++) {
7           data[y][x] = new Complex(image.getRaster().getSample(x, y, 0), 0);
8        }
9     }
10    return data;
11 }
```

Listing 9.4. Converting a `BufferedImage` to `Complex` values.

DFT_TYPE, which will either be DFT_TYPE.FORWARD or DFT_TYPE.INVERSE. A two-dimensional array of Complex type is passed in and a new array of complexes is created and returned.

```
1  public enum DFT_TYPE {FORWARD, INVERSE};
2  public Complex[][] discreteFourierTransform(Complex[][] source,
3                                              DFT_TYPE type) {
4     int width = source[0].length;
5     int height = source.length;
6     int factor = type == DFT_TYPE.FORWARD ? -1 : 1;
7     Complex[][] result = new Complex[height][width];
8     for(int u=0; u<width; u++){
9        for(int v=0; v<height; v++){
10          Complex coefficient = new Complex();
11          for(int x=0; x<width; x++){
12             for(int y=0; y<height; y++) {
13                Complex basis =
14                   Complex.e(1,factor*2*Math.PI*(u*x/width + v*y/height));
15                coefficient = coefficient.plus(source[y][x].times(basis));
16             }
17          }
18          result[v][u] = coefficient;
19       }
20    }
21    return result;
22 }
```

Listing 9.5. Brute force DFT.

The naive implementation of Figure 9.5 has very poor runtime performance due to the fact that computation of every DFT coefficient requires a multiplicative accumulation over every spatial domain sample. More specifically, for an $N \times N$ image this implementation has a runtime complexity of $O(N^4)$, which is so poor that this implementation becomes unusable for even moderately sized images.

This performance can be easily improved by noting that the two-dimensional DFT can be computed by performing a one-dimensional DFT on each row of the source followed by application of the one-dimensional DFT on each column. In other words, the two-dimensional DFT is separable and since the one-dimensional DFT implementation has complexity $O(N^2)$ the use of separability results in a runtime of $O(N^3)$ for a $N \times N$ source.

Fast Fourier transform. One of the most important algorithms ever developed, the fast Fourier transform (FFT), yields an especially efficient implementation of the DFT by clever elimination of redundant computations. The FFT is used in a plethora of domains to solve problems in spectral analysis, audio signal processing, computational chemistry, error correcting codes, and of course image processing. The FFT is so important that it is often implemented in hardware designed specifically for the task.

While various FFT techniques have been discovered, the term FFT typically refers to the Cooley-Tukey algorithm. This technique was discovered and rediscovered by a variety of mathematicians but was popularized by J. W. Cooley and J. W. Tukey in their 1965 publication entitled *An Algorithm for the Machine Calculation of Complex Fourier Series* [Cooley and Tukey 65].

A typical divide and conquer strategy forms the basic premise of the FFT. For the one-dimensional case, an input of length N is divided into two subsections of equal length. The first section contains all of the even-indexed samples of the input while the second section contains all of the odd-indexed samples. The FFT of each of the components is then computed and the results are combined by multiplying the terms by complex roots of unity. Since the input must be divided into two equally sized subsections at every step, the algorithm is limited to handling those inputs where N is a power of 2. Variants on the Cooley-Tukey algorithm have been developed that are able to compute the DFT of a sequence of any length. Implementations that require the input to be a power of 2 are known as radix-2.

The FFT can be formally defined in very concise form. Letting $f(x)$ be a sequence of N samples we denote the sequence of all even-indexed values of $f(x)$ as $e(x)$ and the sequence of all odd-indexed values of $f(x)$ as $o(x)$. In other words $e(x) = (f(0), f(2), \ldots, f(N-2))$ and $o(x) = (f(1), f(3), \ldots, f(N-1))$. We then denote the FFT of $e(x)$ and $o(x)$ as $\mathcal{E}(u)$ and $\mathcal{O}(u)$, respectively. The Fourier coefficients of $f(x)$ are then given by Equation 9.31, where $M = N/2$

and of course $u \in [0, N - 1]$:

$$
\mathcal{F}(u) = \begin{cases} \mathcal{E}(u) + e^{-\frac{j2\pi u}{N}} \mathcal{O}(u) & \text{if } u < M, \\ \\ \mathcal{E}(u - M) - e^{-\frac{j2\pi(u-M)}{N}} \mathcal{O}(u - M) & \text{if } u \geq M. \end{cases} \tag{9.31}
$$

Since the FFT as previously described is naturally recursive it is important to establish a point at which recursion terminates. The recursive base case occurs when the input cannot be divided in half. When $N = 1$ the input cannot be further divided, in which case $\mathcal{F}(0) = f(0)$. In other words, since the first Fourier coefficient is the average of all input samples and since there is a single input sample, the average of all the input samples is that sample.

```
1   public static void fft (boolean forward, Complex[] data) {
2     if (data.length == 1) {
3       return;
4     }
5     int m = data.length/2;
6     double term = −2*Math.PI / data.length;
7
8     Complex[] even = new Complex[m];
9     Complex[] odd = new Complex[m];
10    for(int i=0; i < m; i++) {
11      even[i] = data[i*2];
12      odd[i] = data[i*2+1];
13    }
14
15    fft(forward, even);
16    fft(forward, odd);
17
18    for(int k=0; k<n; k++) {
19      Complex tmp = Complex.e(term *k);
20      if (!forward) tmp = tmp.conjugate();
21
22      data[k] = tmp.times(odd[k]).plus(even[k]);
23      data[k+m] = even[k].minus(tmp.times(odd[k]));
24    }
25
26    if (!forward) {
27      for(int i=0; i<data.length; i++){
28        data[i] = data[i].multiply(.5f);
29      }
30    }
31  }
```

Listing 9.6. FFT.

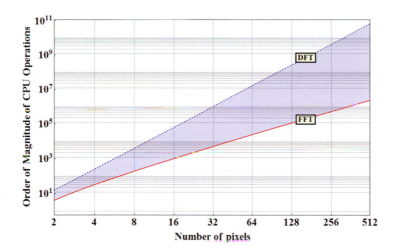

Figure 9.20. FFT complexity.

The forward and inverse FFT transforms are almost identical except for scaling and substitution of the complex conjugate of the exponential term of Equation (9.31). Implementations typically use a single method that performs either the forward or inverse FFT depending on a global flag setting as is shown in the method of Listing 9.6. This method requires that the length of the input array be a power of two where the `forward` flag indicates whether the forward or inverse transform is applied.

Although this implementation is not optimized, it has a runtime performance of $O(N \log N)$ providing a significant performance improvement over the brute force technique. Figure 9.20 shows the relative difference in runtime complexity between the naive DFT implementation and the FFT. The size of a $N \times N$ image is given by the horizontal axis while an approximation to the number of arithmetic operations required for computation is given by the vertical axis. It is important to note that both the vertical and horizontal axes are log compressed. For a 4×4 image the difference between the DFT and FFT is on the order of 10 times while the difference for a 512×512 image is on the order of 10,000 times.

9.5   Wavelets

Similar to the Fourier series, wavelets are a class of functions that decompose a signal into a set of basis functions and that have properties superior to the the DFT for some applications. In wavelet transformations, the basis functions are all scaled and translated copies of a waveform known as the *mother wavelet*. While a rigorous overview of wavelet theory is beyond the scope of this text, we will here present a very brief overview of wavelets as used in image processing applications.

The discrete wavelet transform (DWT) is used to convert a finite-length discrete sequence into a sequence of wavelet coefficients using a mother wavelet to define the basis functions. The DWT is a recursive process where at each stage of the transform the input is scaled and a set of average (or low frequency) coefficients and detail (or high frequency) coefficients is generated. Each subsequent stage operates only on the low frequency terms; thus generating a pyramid of increasingly coarse approximations of original sequence in addition to the high frequency content that can be used to reconstruct the original.

Consider a digital grayscale image with eight columns where a single row of the image contains the samples (15, 9, 13, 15, 11, 9, 7, 9). We first scale the input by a factor of two by averaging adjacent samples. The low frequency component is then given as $((15+9)/2, (13+15)/2, (11+9)/2, (7+9)/2)$ or $(12, 14, 10, 8)$. The high frequency coefficients allow us to reconstruct the original signal from the low frequency component. Since the average value of the two samples lies at the midpoint between them, it is sufficient to record the difference between the average value and one of the samples. In this case, the detail coefficients are given as $(15-12, 13-14, 11-10, 7-8)$ or $(3, -1, 1, -1)$. This process is repeated for the low frequency component until no more subdivisions can be made such that the DWT coefficients are finally given as $(11, 2, -1, 1, 3, -1, 1, -1)$. Note that the first DWT coefficient is the coarsest approximation to the entire original sequence, the average of all values. Figure 9.21 shows how this process generates a sequence of increasingly coarse versions of the original sample.

The left-hand segments at each level correspond to the low pass coefficients while the right-hand segments correspond to the detail coefficients. Wavelets provide excellent support for multi-resolution analysis since the coefficients contain easily reconstructed scalings at various resolutions of the original image. If computational speed is desired then the analysis can be performed on one of the coarser scalings while more accurate analysis can take place at the finer grained reconstructions. For each level in the transform, the elements of the preceding

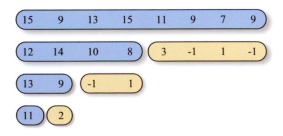

Figure 9.21. Haar wavelet example.

(a) Source image.　　　　　　　　　　　　　(b) Haar coefficients.

Figure 9.22. Haar wavelet coefficients.

level can be reconstructed by properly summing the detail coefficients with the average coefficients. The example given in the preceding paragraphs is based on the Haar wavelet, which is the simplest of the mother wavelets. The most commonly used wavelets in image processing are related to the Daubechies, Coiflet, and the Mexican hat wavelet families.

The two-dimensional wavelet transform is separable and hence can be easily performed by repeated application of the one dimensional DWT in the same way as described for the DFT. Figure 9.22 gives an illustration of the Haar wavelet coefficients computed only to the second level. Note that the scaled original is to the upper left quadrant while the high frequency details span the remaining three quadrants. The details for horizontal change are given in the upper right quadrant, the details for vertical change given in the lower left quadrant, and the details governing the derivative of these occurs in the lower right quadrant. A brief discussion of the DWT as it is used in image compression is found in Chapter 10.

9.6　Exercises

1. Give the DCT coefficients for the four sample sequence (18, 36, 20, 12).

2. Give the DFT coefficients for the four sample sequence (18, 36, 20, 12).

3. Design and implement an abstract class named `WindowingOp` that performs windowing as described in Section 9.4.3. Design subclasses that are able to perform the windowing filter using either the Bartlett, Hanning, Blackman, or Hamming technique.

4. Use Pixel Jelly to generate a 256×256 checkerboard pattern where the checkerboard tiles are of size 32×32. Use the DFT dialog to apply a low pass filter and record the images that result as the cutoff is increased from 0 to 20 in step sizes of 2. Discuss the results. Repeat this process with an image made of ovals rather than checkered tiles.

5. Use Pixel Jelly to load a binary image of some power-of-two dimension. Generate an image of the DFT magnitude as the image is (a) rotated by $45°$, (b) scaled down by 50%, and (c) translated to the edge of the image. Discuss the results and show how they are related to the properties of the DFT.

6. Use Pixel Jelly to load an image and produce a 3×3 tiled copy of this image. Compare and contrast the non-windowed DFT magnitude spectrum with the windowed DFT magnitude spectrum.

7. Consider the sequence given by (83, 75, 109, 45, 39, 42, 55, 82). Give the Haar wavelet DWT coefficients.

8. Consider using the Haar wavelet to generate the DWT coefficients given by $(62, -38, 5, -1, 3, 1, -1, -10)$. Reconstruct the original sequence.

Artwork

Figure 9.6. *The Portrait of Carolina Zucchi* by Francesco Hayez (1791–1882). Francesco Hayez was an Italian painter and one of the leading 19th century Romanticists. Hayez is most known for his portraiture and large-scale historical paintings. Hayez was the youngest of five sons and came from a relatively poor family. He displayed talent and interest in art at a young age so his uncle secured him an internship working under an art restorer. He later trained under Francisco Magiotto and was admitted to the New Academy of Fine Arts in 1806. Hayez did not sign or date his work and he often painted the same scene with little or no variation several times. Verifying the authenticity and time line of his career is thus made more difficult as a consequence.

Figure 9.7. *Suprematism* by Kazimir Malevich (1878–1935). Kazimir Malevich was a Russian painter who pioneered geometric abstract art and was the originator of the Suprematist movement, which focused on fundamental geometric forms. His paintings include a black circle on white canvas in addition to a black square on white canvas. Malevich felt that the simplicity of these forms were symbolic of a new beginning that involved a sort of mystic transcendence over the normal sensory experiences of life. Figure 9.7 shows his 1916 work, which goes by the title of the movement he began. It seemed appropriate to use his work to illustrate the spectral properties of shape.

Figure 9.8. *Suprematism* by Kazimir Malevich (1878–1935). See the entry for Figure 9.7.

Figure 9.11. *Daniel Chodowiecki on the Jannowitzbrucke* by Adolph Friedrich Erdmann von Menzel (1815–1905). Adoph Menzel was a German artist noted for drawings, etchings, and paintings. His first commercial success came in 1883 when a collection of pen and ink drawings was published. Over the three-year period of 1839–1842 he produced some 400 illustrations for a book entitled *The History of Frederick the Great* by Franz Kugler. The detail of Figure 9.11 is a portrait of the Polish-German illustrator Daniel Chodoweiecki standing on a Berlin bridge.

Figure 9.13. *Daniel Chodowiecki on the Jannowitzbrucke* by Adolph Friedrich Erdmann von Menzel (1815–1905). See the previous entry for a brief biographical sketch of the artist.

Figures 9.14 and 9.17. *Suprematism* by Kazimir Malevich (1878–1935). See the entry for Figure 9.7.

Figure 9.19. *Still Life with Salmon* by Édouard Manet (1832–1883). Édouard Manet was a French painter who produced some of the most popular works of the 19th century. Manet was born into a wealthy family and later, after twice failing the entrance exam for the French Navy, was allowed to pursue an education in the arts. Manet studied under Thomas Couture and developed his skill by painting copies of the old masters in the Louvre. His work propelled the transition from Realism to Impressionism, fueled in large part by controversy that surrounded *The Luncheon on the Grass* and *Olympia*. Both of these works appear to feature prostitutes, which were common in his time but seen as inappropriate subjects for public display. See Figure 9.19 on page 244.

Figure 9.22. *The Kiss* by Gustav Klimt (1862–1918). Klimt was born in Baumgarten near Vienna. His early training was at the Vienna School of Arts and Crafts where his work was disciplined and academic. He later helped to found the Vienna Secession, which was a group of artists dedicated to promoting the work of the young and unconventional, regardless of style. Although Klimt was well known and prosperous during his lifetime, his work was also highly contro-versial due to its sexually charged themes. Although the image of Figure 9.22 is perhaps his best known work, the work entitled *Adele Bloch-Bauer I* sold for US$135 million in 2006, making it one of the most expensive works of art in the world.

Image Compression 10

10.1 Overview

The primary goal of image compression is to minimize the memory footprint of image data so that storage and transmission times are minimized. Producing compact image data is essential in many image processing systems since storage capacity can be limited, as is the case with digital cameras, or costly, as is the case when creating large warehouses of image data. Transmission of image data is also a central concern in many image processing systems. Recent studies of web use, for example, have estimated that images and video account for approximately 85% of all Internet traffic. Reducing the memory footprint of image data will correspondingly reduce Internet bandwidth consumption. More importantly, however, since most web documents contain image data it is vital that the image data be transferred over the network within a reasonable time frame. Reducing the memory footprint has the significant advantage of speeding delivery of web content to the end user.

Image compression works by identifying and eliminating redundant, or duplicated, data from a source image. There are three main sources of redundancy in image compression. The first is known as interpixel redundancy, which recognizes that pixels that are in close proximity within an image are generally related to each other. A compression technique that reduces memory by recognizing *some* relationship between pixels based on their proximity is an attempt to eliminate interpixel redundancy. Run length encoding, constant area coding, and JPEG encoding seek to eliminate this source of unnecessary data.

The second source of redundancy is known as psycho-visual redundancy. Since the human visual system does not perceive all visible information with equal sensitivity we understand that some visual information is less important than others. Image compression systems will simply eliminate information that is deemed to be unimportant in terms of human perception. Note that since reducing psycho-visual redundancy results in information loss, the process is not

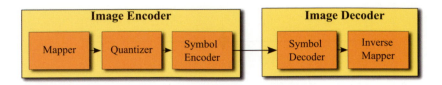

Figure 10.1. General model of (a) an image encoder and (b) an image decoder.

reversible and hence the compressed image suffers a loss of quality. The lossy constant area coding technique described later in this section is an example of reducing psycho-visual redundancy.

And finally, coding redundancy refers to the way in which bits are used to represent the image information. If image samples are stored in such a way that more bits are required than necessary, then there is redundancy in the encoding scheme. Although coding redundancy will not be directly addressed in this text it should be noted that general purpose compression techniques such as huffman or arithmetic encoding are effective at reducing coding redundancy.

A software module that compresses image data is an image encoder while a software module that decompresses image data is an image decoder. An image encoder generally consists of three primary components, each of which seeks to address one of the three source of redundancy. Figure 10.1 shows a block diagram of the main components of an image encoder. The mapper transforms an image into a form, or representation, such that interpixel redundancy is either eliminated or reduced. Some compression techniques convert spatial domain information into the frequency domain, for example, and this transformation is considered to be part of the mapper. The quantizer changes the information produced by the mapper into a discrete set of values and may even truncate data such that some information is lost in the process. The quantizer typically eliminates or reduces psycho-visual redundancy. Symbol encoding is then performed on the resulting data stream in order to reduce coding redundancy. An image decoder will then decode the symbol data stream and perform an inverse mapping operation to obtain the resulting image.

A wide variety of image compression techniques are in common use where PNG, JPEG, and GIF are among the most popular. The effectiveness of an image processing method can be assessed by determining (1) the computational complexity of the encoding phase, (2) the computational complexity of the decoding phase, (3) the compression ratio, and (4) the visual quality of the result.

This text will not consider the computational complexity of either the encoding or decoding process other than to note that efficient encoding of image data is not generally of concern but that efficient decoding is often crucial. Consider, for example, a web-based streaming video service. The web server may take a long time to compress a video stream in order to produce highly compact video data that can be streamed to clients on low bandwidth connections. But note that while compression can be performed at slow speeds, the client must be able to decode the compressed data stream in real time so as to render and view the resulting video.

Compression ratio serves as the primary measure of a compression technique's effectiveness. Compression ratio is a measure of the number of bits that can be eliminated from an uncompressed representation of a source image. Let $N1$ be the total number of bits required to store an uncompressed (raw) source image and let $N2$ be the total number of bits required to store the compressed data. The compression ratio C_r is then defined as the ratio of $N1$ to $N2$, as shown in Equation (10.1). Larger compression ratios indicate more effective compression while a compression ratio lower than unity indicates that the compressed image consumes more memory than the raw representation:

$$C_r = \frac{N_1}{N_2}.$$ (10.1)

The compression ratio is sometimes referred to as the *relative* compression ratio since it compares the memory footprint of one representation to another. Throughout this text, the raw format will serve as the baseline against which the relative compression ratios will be compared. The raw format is assumed to be any format that uses 8 bits per sample, in the case of color and grayscale images, or 1 bit in the case of binary images. The compression ratio is commonly simplified by using the notation N1:N2. If an image in raw format consumes 5 Megabytes of memory while in compressed form it consumes 1 Megabyte, the relative compression ratio can be given as 5:1, which should be read as *5 to 1*.

The savings ratio is related to the compression ratio and is a measure of the amount of redundancy between two representations. The savings ratio is defined in Equation (10.2). The savings ratio is a percentage of how much data in the original image was eliminated to obtain the compressed image. If, for example, a 5 Megabyte image is compressed into a 1 Megabyte image, the savings ratio is defined as (5-1)/5 or 4/5 or 80%. This ratio indicates that 80% of the uncompressed data has been eliminated in the compressed encoding. This is not to imply that the compressed image is of inferior quality (although it may be) but simply that 80% of the original images data was in some sense redundant and could therefore be eliminated. Higher ratios indicate more effective compression while negative ratios are possible and indicate that the compressed image exceeds the memory footprint of the original.

$$S_r = \frac{(N_1 - N_2)}{N_1}.$$ (10.2)

Root mean squared (RMS) error is a generally accepted way of measuring the quality of a compressed image as compared with the uncompressed original.

RMS error is a measure of the difference between two same-sized images and is not related to the memory footprint of an image. Assume that a $W \times H$ image I having B bands is compressed into image I'. The root mean square error, e_{rms}, is then given in Equation (10.3):

$$e_{rms} = \sqrt{\frac{\sum\limits_{x=0}^{W-1}\sum\limits_{y=0}^{H-1}\sum\limits_{b=0}^{B-1}(I'(x,y)_b - I(x,y)_b)^2}{W \cdot H \cdot B}}. \qquad (10.3)$$

The RMS error is a measure of the average sample error between two images. This can be seen by recognizing that the total number of samples in the image, $W \cdot H \cdot B$, occurs in the denominator and that the numerator sums the squares of the errors between every pair of corresponding samples in the two images. Since RMS is a measure of error, small RMS measures indicate high-fidelity compression techniques while techniques that produce high RMS values are of lower fidelity. If, for example, every sample of the compressed image is

(a) RAW   (b) JPEG (low quality)   (c) PNG
1,028,054 bytes.   5,710 bytes   708,538 bytes
                                  $C_r = 99.4\%$ $C_r = 31.1\%$
                                  $e_{rms} = 20.4.$ $e_{rms} = 0.0.$

Figure 10.2. Compression and bandwidth.

identical to the corresponding sample of the original, the RMS error is 0. Any compression technique that can be characterized by a high compression ratio and a low expected RMS value is to be preferred, all else being equal.

An image compression technique can be broadly classified as either lossy or lossless. A lossless technique is one which always produces a compressed image with an e_{rms} of 0 relative to the source. A lossy technique, however, generates a compressed image that is not identical to the source. Lossy techniques are typically able to achieve greater compression ratios by sacrificing the quality of the result. Figure 10.2 gives an illustration of the typical tradeoffs between compression ratio and fidelity. In this example a 24-bit per pixel 513×668 image consumes $24 \cdot 513 \cdot 668 = 1,028,054$ bytes without any compression. JPEG compression reduces the memory footprint to 6,923 bytes but significantly reduces the image quality while PNG compression reduces the memory footprint to 708,538 bytes without any reduction in quality. The JPEG technique achieved a compression ratio of 99.4% while the PNG approach delivered a 33.1% compression ratio. It should be noted that JPEG can be controlled so as to provide much higher quality results but with a corresponding loss of compression.

10.2   Run Length Coding

Run length encoding is a lossless encoding scheme in which a sequence of same-colored pixels is stored as a single value rather than recording each pixel individually. Consider, for example, a binary image in which each pixel is represented by a single bit that is either 0 (black) or 1 (white). One row of the image may contain the 32-pixel sequence

$$11111111110001111111111111111111$$

This row contains three runs: 10 white pixels followed by 3 black followed by 19 white. This information can be completely encoded by the three byte sequence {10, 3, 19} by assuming that the data begins with white runs. The original representation consumes 32 bits of memory while the run length-encoded representation consumes 24 bits of memory if we assume that 8-bit bytes are used for each run. This results in a compression ratio of 4:3 and a savings ratio of

$$S_r = \frac{(N_1 - N_2)}{N_1} = \frac{(32 - 24)}{32} = \frac{8}{32} = 25\%.$$

Figure 10.3 illustrates how run length encoding can represent binary image data. In the illustration, one row of an image of the character R is encoded. The row contains a total of 88 bits (11 bytes) and five runs. This information is losslessly encoded with the 5-byte sequence {20, 7, 22, 18, 21}, which consumes

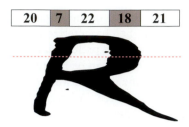

Figure 10.3. Run length encoding.

a total of 40 bits and results in a compression ratio of 11:5 and a savings ratio of $S_r \simeq 54\%$.

Listing 10.1 gives an algorithm for run length encoding binary images where the encoder assumes that the first run encoded on each row is composed of white pixels. This is a reasonable assumption since most monochrome images are of black foreground objects on white background and the first run will almost always be comprised of white pixels. If the first pixel on any row is black, the length of the first white run is zero. Run length encoding is most effective for images where long runs of identical pixels are common. Most types of monochrome images (e.g., line drawings and scanned text), or palette-based iconic images are amenable to run length encoding while continuous tone images can not generally be effectively compressed with this technique since long runs of equal samples are rare.

Implementation of a run length encoder for binary images brings up an important technical detail that must be addressed. If run lengths are encoded with 8-bit bytes the question of how to represent runs that exceed 255 in length must be answered since bytes can only range in value between 0 and 255. A straightforward solution is to encode a single run length as a multi-byte sequence. A

```
1  runlengthEncode(Image source, OutputStream OS)
2    writeTheHeader(source)
3    for every row R of the source image
4      boolean isWhite = true;
5      while there are more runs on row R
6        L = length of next run
7        write L to OS
8        isWhite = !isWhite;
```

Listing 10.1. Run length encoding algorithm.

single run of 783 pixels, for example, could be encoded as the 4-byte sequence {255, 255, 255, 18}, where the byte value 255 signifies that this run is 255 pixels longer and the run is not over yet and any byte b other than 255 signifies that this run is B pixels longer and the run is now over. Given this interpretation, the encoded run length {255, 255, 255, 18} can be decoded as 255 + 255 + 255 + 18 = 783 pixels in length. Under this encoding scheme a run of length 255 is encoded as the two byte sequence {255, 0}.

While the algorithm of Listing 10.1 shows how to run length encode a binary image, consider the problem of encoding a grayscale or color image. For an 8-bit grayscale image there are more than two possible sample values and hence a run must somehow be encoded as a two-element pair where the first element is the value of the run and the second is the length of the run. When encoding a row profile containing the samples {100, 100, 103, 104, 110, 110, 110}, for example, the run length representation would be {(100,2), (103,1), (104,1), (110,3)}. Since storage is required for the *value* of the run in addition to the length of the run, the effectiveness of run length encoding is reduced. This example also highlights another problem with run length encoding of grayscale images since it is extremely unlikely that a grayscale image contains runs of any significant length.

Consider, however, another approach to run length encoding a grayscale image. Since run length encoding a binary image is generally effective we can view a grayscale image as a collection of binary images that can be encoded using the technique of Listing 10.1. Just as a single 24 bpp image can be split into three 8-bit grayscale images, an 8-bit grayscale sample can be split into eight 1-bit or binary bands. All grayscale and color images can be decomposed into a set of binary images by viewing each sample as eight individual bits that form eight separate bands. Figure 10.4 illustrates how a 4-bit grayscale image can be decomposed into four separate monochrome images by bit plane slicing. A 3×2 4-bit grayscale image is shown in part (b) where the samples are displayed in binary notation. The least significant bit of each sample forms the 0th bit plane P_0 while the most significant bit of each sample forms the 3rd bit plane P_3, which are shown as the four binary images of parts (b) through (e). In this illustration, a 1 bit is rendered as a white pixel while a 0 bit is rendered as black.

Note that the 7th bit plane of an 8-bit grayscale image is equivalent to the result of thresholding with a cutoff value of 127 since the 7th bit is 1, or white, for those samples exceeding 127 and is 0, or black, for those samples with a value

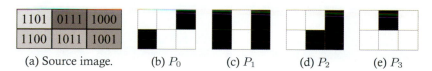

| (a) Source image. | (b) P_0 | (c) P_1 | (d) P_2 | (e) P_3 |

Figure 10.4. The bit planes of a 4 bpp source image.

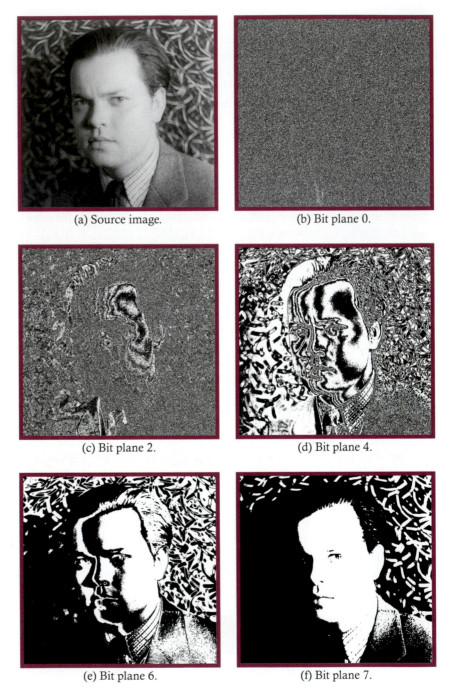

(a) Source image.

(b) Bit plane 0.

(c) Bit plane 2.

(d) Bit plane 4.

(e) Bit plane 6.

(f) Bit plane 7.

Figure 10.5. Bit planes as binary images.

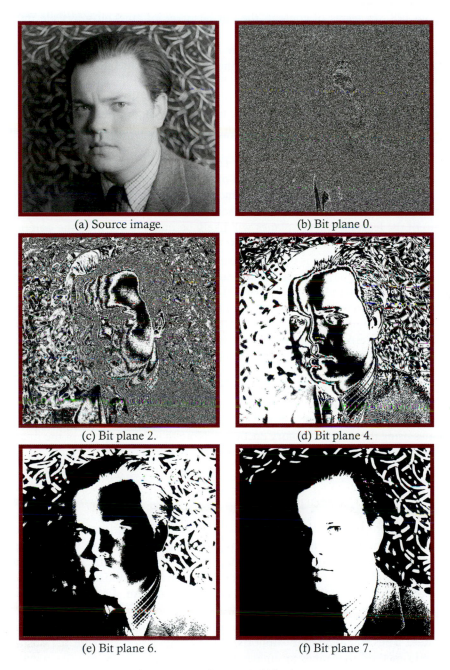

(a) Source image.

(b) Bit plane 0.

(c) Bit plane 2.

(d) Bit plane 4.

(e) Bit plane 6.

(f) Bit plane 7.

Figure 10.6. Gray coded bit planes.

below 128. In terms of content, the 7th bit plane of an 8 bit grayscale image consists of the most significant bits of each sample and hence more accurately represents the structure of the grayscale image than the other planes. Since the least significant bit plane consists of the least significant bits of each sample, the resulting bit planes contain relatively little information and are sometimes considered to be completely irrelevant or noise. Figure 10.5 shows a number of bit planes constructed from a grayscale image Note that in this figure, the more significant planes contain more structural information than the less significant bit planes.

Understanding that a grayscale or color image can be decomposed into a sequence of binary images allow us to modify the algorithm of Listing 10.1 by using bitplane slicing. We repeatedly encode the individual bit planes of an image using the binary image encoder where we understand that the compression ratio of the least significant planes will likely be much lower than the compression ratio of the most significant planes.

Run length encoding is most effective for images having long runs. Consider a grayscale image having adjacent samples of 127 and 128. In binary form these two samples are 01111111 and 10000000. While the two samples are a single intensity level in difference, all eight bits of the two samples are different and hence any run on any plane will be broken at the junction of these two samples. Gray coding is an alternate binary representation that can be useful for extending runs across bit planes. The reflected binary code, also known as gray code after the discoverer Frank Gray, is a binary numeral system where two successive values differ in only one bit. A grayscale image can be decomposed into 8 bit planes using a gray coding scheme in order to increase the length of runs across all of the bit planes. Figure 10.6 shows how a grayscale image can be decomposed into bit planes using the reflected binary encoding. Note that the structure of the images is more coherent at every plane than the corresponding planes of Figure 10.5.

10.2.1   Implementation

Image compression requires a great deal of attention to low level data types and careful attention to robust design techniques. In this textbook we will develop a general framework that will support image compression by first describing two classes: one class for compressing and encoding an image into an output stream, and another for decoding an image from an input stream. Our framework will closely mirror the standard Java framework.

Listing 10.2 gives a listing of the ImageEncoder and ImageDecoder abstract classes. The ImageEncoder supports image compression through the abstract encode method that takes an image as input and encodes it into the specified data output object. The DataOutput class is an interface that is typically implemented as an output stream. An OutputStream represents any object to which

```
1  public abstract class ImageEncoder {
2    public void writeHeader(BufferedImage source, DataOutput output)
3        throws IOException {
4      output.writeUTF(getMagicWord());
5      output.writeShort((short) source.getWidth());
6      output.writeShort((short) source.getHeight());
7      output.writeInt(source.getType());
8    }
9
10   public void encode(BufferedImage source, File f) throws IOException {
11     FileOutputStream fout = new FileOutputStream(f);
12     encode(source, fout);
13     fout.close();
14   }
15
16   public abstract String getMagicWord();
17   public abstract void encode(BufferedImage image, OutputStream os)
18       throws IOException;
19 }
20
21 public abstract class ImageDecoder {
22   public boolean canDecode(File file) {
23     try {
24       DataInputStream fin = new DataInputStream(new FileInputStream(file));
25       String mw = fin.readUTF();
26       fin.close();
27       return (mw != null && mw.equals(getMagicWord()));
28     } catch(Exception e) {
29       return false;
30     }
31   }
32
33   public BufferedImage decode(File f) throws IOException {
34     FileInputStream fin = new FileInputStream(f);
35     BufferedImage result = decode(fin);
36     fin.close();
37     return result;
38   }
39
40   public abstract String getMagicWord();
41   public abstract BufferedImage decode(InputStream is) throws IOException;
42 }
```

Listing 10.2. ImageEncoder and Decoder.

data can be written, generally one byte at a time. While an output stream is often associated with a file, it may be associated with a network connection or any other device to which data can be written. The `ImageDecoder` provides the inverse service through the abstract `decode` method whereby data is read from an input stream which is then used to construct an image. Both the encode and decode methods are overloaded such that clients may encode or decode directly to files. In both cases, the method simply attaches a stream to the specified file and the calls the coding method to perform the actual encoding or decoding.

Most image file formats begin with a *magic number* or, more descriptively, a magic word. The magic number is a sequence of bytes that identifies the file format and hence how the file should be decoded. In our implementation, the magic word is simply text such that each encoder and decoder can be uniquely identified. The `writeHeader` method of line 2 writes information necessary for recovering an encoded image. The magic word followed by the width, height, and the type of image is written into the stream. The *type* of image is the value returned by the `getType` method of the buffered image class. The `encode` method will initially call the `writeHeader` method and then generate the appropriate encoding. An `ImageDecoder` can open a file and quickly see if it can be decoded by checking the magic word stored in the file header. This is accomplished via the `canDecode` method on line 22. The `decode` method reads necessary header information and then begins decoding the image data.

Observant readers will note that the header method accepts a `DataOutput` object rather than an output stream. A `DataOutput` object is able to write byte-oriented data in a manner similar to output streams but is also able to write other types of objects, such as all of the primitive data types and Strings. Since the header consists of ints, shorts, and Strings, the `DataOutput` object is useful there. A `DataOutputStream` is a `DataOutput` implementation.

To implement a run length encoder we must write concrete subclasses of both the encoder and decoder. Listing 10.3 gives a complete implementation of a run length encoder. The encode method accepts an output stream and transforms it a `DataOutputStream` so that primitive data can be written to it rather than just individual bytes. The header consists of the magic word "RLE" to indicate that this image has been run length encoded. This header information is necessary for the decoder so that it knows how to decode the stream and the type of image that it should return.

Each row of each bitplane is then encoded using a run length encoding technique. The central method is the `getRun` method, which computes the length of the next run on the given bitplane of the specified band.

The length of the run is then written to the output stream via the `writeRun` method, which writes as many bytes as necessary in order to encode the run.

```java
class RunLengthEncoder implements ImageEncoder {
  private void writeRun(int length, DataOutputStream os) throws IOException {
    while(length >= 255) {
      os.write(255);
      length -= 255;
    }
    os.write(length);
  }

  public String getMagicWord() {
    return "RLE";
  }

  private boolean getBit(int sample, int bit) {
    return (sample & (0x1 << bit)) != 0;
  }

  private int getRun(BufferedImage source, int column, int row,
                     int band, int bit, boolean isWhite) {
    int result = 0;
    while(column < source.getWidth() &&
      getBit(source.getRaster().getSample(column,row,band),bit)==isWhite) {
        column++;
        result++;
    }
    return result;
  }

  public void encode(BufferedImage source, OutputStream out)
      throws IOException {
    DataOutputStream output = new DataOutputStream(out);

    writeHeader(source, output);
    for(int band=0; band<source.getSampleModel().getNumBands(); band++) {
      for(int bit=0; bit<source.getSampleModel().getSampleSize(band); bit++) {
        for(int y=0; y<source.getHeight(); y++) {
          boolean isWhite = true;
          int pixelsEncoded = 0;
          while(pixelsEncoded < source.getWidth()) {
            int length = getRun(source, pixelsEncoded, y, band, bit, isWhite);
            isWhite = !isWhite;
            writeRun(length, output);
            pixelsEncoded += length;
          }
        }
      }
    }
  }
}
```

Listing 10.3. RunLengthEncoder.

10.3   Hierarchical Coding

In run length encoding each datum represents a one-dimensional run of pixels. Constant area coding (CAC) is a two-dimensional extension of this idea where entire rectangular regions of pixels are stored as a single unit. While run-length encoding is a lossless compression technique, constant area coding can be either lossy or lossless depending upon the specific implementation.

Consider encoding a $W \times H$ binary image using a CAC technique. If the entire image is white we simply output white and are done with the encoding. If the entire image is black we output black and are also done encoding. If, however, the image is neither all white nor all black the image is divided into four equally sized regions by splitting the image in half both vertically and horizontally and each of these four regions is then recursively encoded following the same procedure. The process recurses only for non-homogeneous regions and will always terminate since at some point the region will be no larger than a single pixel which will be either all white or all black.

The division of an image region into four equal-sized subregions is often represented through a region quadtree. A region quadtree is a data structure that partitions two-dimensional rectangular regions into subregions as described in the preceding paragraph. Each node of a quadtree contains either zero or four children and represents a rectangular region of two dimensional space. Any node having no children is a leaf (or terminal) node while any node having four children is an internal node. Leaf nodes represent rectangular regions that are completely filled by a single color while internal nodes represent regions containing variously colored quadrants as described by their four children.

Figure 10.7 provides an illustration of quadtrees. A $W \times H$ rectangular region R is divided into four quadrants, each of which is a leaf in this example. While the image itself may have large dimension the corresponding quadtree has exactly five nodes and is able to completely encode the entire image in only a few bytes. The quadtree representation of R is visualized in part (b) of Figure 10.7, where leaves are drawn as rectangles filled with the leaves colors while internal nodes are drawn as ovals.

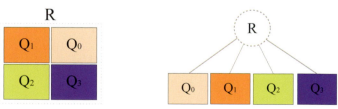

(a) Region R partitioned into quadrants.   (b) Quadtree representation of R.

Figure 10.7. An illustration showing the quadtree representation of an image region.

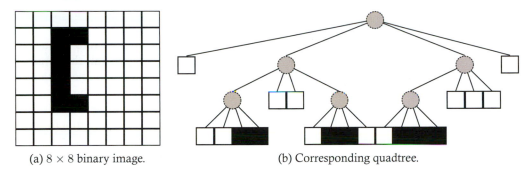

(a) 8×8 binary image.

(b) Corresponding quadtree.

Figure 10.8. Example of CAC.

Figure 10.8 provides a concrete example of how an 8×8 monochrome image can be represented by a region quadtree. Since the 8×8 image is neither all black nor all white it must be subdivided into four 4×4 quadrants that become the children of the root node. Since quadrants Q_0 and Q_3 are all white, we tag those regions as white and they are not subdivided further. Regions Q_1 and Q_2 are not all white and hence they are further subdivided as seen in Figure 10.8(b). Each level of the tree corresponds to a region that is half the width and height of its upper level. Since the image is 8×8, the top node represents the entire 8×8 region while the four second level nodes each represent a 4×4 region, the eight third level nodes each represents a 2×2 region and the twelve lowest-level nodes each represent a 1×1 region of the image.

Having constructed the quadtree, the quadtree must be serialized, or written to an output stream bit by bit. A quadtree can be serialized by traversing the tree in pre-order fashion and writing a single bit for every node visited. For every visited node we write a 0 bit if the region represented by the node contains at least one black pixel and a 1 bit otherwise. Although it is possible to construct binary images where the quadtree representation consumes more memory than the RAW format, most binary images can be effectively compressed using CAC encoding. Line drawings, scanned text, and palette-based iconic images are particularly well suited for this scheme.

When the quadtree of Figure 10.8 is serialized in this fashion, it results in the bit sequence 0100110011010010010001111. Since the original image is binary each pixel is a single bit and hence the entire image consumes 64 bits of memory in raw form. The serialized quadtree consumes 25 bits of memory, which results in a compression ratio of 61% as given by

$$C_r = \frac{(64 - 25)}{64} = 61\%.$$

10.3.1 Implementation

When implementing a CAC encoder an image is first converted into its quadtree representation after which the tree is then serialized. In general, the serialized quadtree will be more compact than the raw image data if the image contains sufficient regions of identically valued pixels. CAC encoding of binary images is not only effective but is also easily implemented. Our implementation will *implicitly* construct the quadtree through a sequence of recursive function calls and hence there is no need for an explicit tree data type. Each function call represents a single tree node and the series of calls effectively traverses, in a pre-order fashion, the quadtree representation.

Consider the low-level details of encoding a binary image. The source code of Listing 10.4 gives a nearly complete implementation of CAC encoding for binary source images. Each call to the `write` method corresponds to visiting a node in the quadtree covering the region that has (x, y) as its upper-left coordinate and is w pixels wide and h pixels tall. If either the width or height of the region is zero or less then the region contains no pixels and there is nothing to encode as is the case on line 4. If the region is all white, as indicated on line 6, a 1 bit is written. Otherwise the region has at least one black pixel so a 0 bit is written and the region is subdivided into four quadrants and each quadrant is then encoded by recursively calling the `write` method, as shown by the condition of line 8.

The `MemoryCacheImageOutputStream` is a standard but nonetheless specialized Java class in the `javax.imageio` package. Objects of this class are created

```
1  private void write(BufferedImage source,
2                     MemoryCacheImageOutputStream os,
3                     int x, int y, int w, int h) throws IOException {
4    if(w <= 0 || h <= 0) return;
5
6    if(isAllWhite(source,x,y,w,h)) {
7      os.writeBit(1);
8    } else {
9      os.writeBit(0);
10     write(source, os, x + w/2, y, w-w/2, h/2);        // quadrant 0
11     write(source, os, x, y, w/2, h/2);                // quadrant 1
12     write(source, os, x, y + h/2, w/2, h - h/2);      // quadrant 2
13     write(source, os, x + w/2, y + h/2, w-w/2, h-h/2);// quadrant 3
14   }
15 }
```

Listing 10.4. CAC implementation.

```
1   public class BinaryCACEncoder extends ImageEncoder {
2     public String getMagicWord() {
3       return "BCA";
4     }
5
6     public void encode(BufferedImage source, OutputStream os)
7                                   throws IOException {
8       MemoryCacheImageOutputStream output =
9         new MemoryCacheImageOutputStream(os);
10      writeHeader(source, output);
11      write(source, output, 0, 0, source.getWidth(), source.getHeight());
12      output.close();
13    }
14
15    private void write(...) { ... }
16    private boolean isAllWhite(...) { ... }
17  }
```

Listing 10.5. BinaryCACEncoder.

by attaching them to some output device to which data is written. The main virtue of a memory cached output stream is its ability to write individual bits in addition to bytes. The name of the class derives from the fact that the output is buffered up in internal memory (or cached in memory) prior to being flushed to the attached output stream.

To complete our implementation we must write a concrete subclass of ImageEncoder such as outlined in Listing 10.5. The magic word is chosen as "BCA" for binary constant area encoding. The encoding method creates a memory cached output stream, an implementation of the DataOutput interface, by attaching it to the client-supplied output stream, writing the header information to the stream, and then calling the write method specifying that the entire image region should be written. The region is defined by the location of the upper left pixel $(0, 0)$ where the width and height are given as the image dimensions.

10.3.2   Encoding Grayscale and Color Images

CAC encoding of grayscale and color images poses more difficulties than the encoding of binary images. Continuous tone images are not likely to contain large regions of precisely identical colors and hence will generally subdivide to regions of one pixel in size. In addition, rather than having only two possible colors there are 256 possible tones for a grayscale image and hence we must store, for each leaf node in the quadtree, an entire byte designating the color of that node rather than the single bit required for binary images.

The algorithm of Listing 10.4 for encoding binary images can, however, be effectively extended to encode grayscale and color images in two different fashions. The first is to view grayscale images as a collection of bit planes and to encode each bit plane using the monochrome technique described above. While this results in a lossless encoding scheme, it is generally the case that the compression ratio will be low since little, if any, compression will occur in the less significant bit-planes and thus negate any compression obtained in the more significant bit planes.

The second technique for CAC encoding grayscale and color images is a lossy technique. Since continuous tone images are not likely to contain large homogeneous regions of identically valued samples, CAC compression can be made effective by encoding regions of *similar* samples *as if they were identical*. In such a scheme, each leaf node of the quadtree contains the average value of the samples over the region. Such an approximation is reasonable if the samples in the region are largely similar. At each stage of lossy CAC compression a region must either be divided into quadrants or left intact and encoded as a whole. If the samples in a region are sufficiently similar we can approximate the entire region as a homogeneous whole by recording the average sample value of all samples in the region.

Rather than dividing a region into quadrants if there is any one pixel that is different than the others, we only divide a region into quadrants if the pixels are not sufficiently similar. Similarity can be measured by the standard deviation such that a larger standard deviation indicates a larger difference in the samples and a smaller standard deviation indicates a smaller difference in the samples. The standard deviation is related to the e_{rms} error given in Equation (10.3), where the mean or average value of the samples in the region serves as the baseline. Equation (10.4) shows how to compute the standard deviation, denoted as σ, for all samples in a $W \times H$ subregion of a grayscale image where the average value of all samples in the subregion is denoted as $\bar{\mu}_{W,H}$.

$$\sigma = \sqrt{\frac{\sum\limits_{x=0}^{W-1} \sum\limits_{y=0}^{H-1} (I(x,y) - \bar{\mu}_{W,H})^2}{W \cdot H}}. \tag{10.4}$$

Lossy CAC encoding will first compute the standard deviation of the samples in a region and divide the region if the standard deviation is larger than some threshold setting. The larger the threshold, the more the image is compressed but

10	9	11	10
10	9	15	9
72	72	12	9
83	78	83	11

10	10	11	11
10	10	11	11
76	76	12	9
76	76	83	11

0	1	0	1
0	1	-4	2
4	4	0	0
-7	-2	0	0

(a) Source image.   (b) Encoded.   (c) Difference.

Figure 10.9. Numeric example of grayscale CAC encoding.

at the cost of lower fidelity. Higher threshold settings signify that regions are allowed to contain greater error, which generally results in low fidelity but with greater compression ratios. Lower threshold settings indicate that regions are not allowed to contain much difference and this results in lower compression ratios but higher fidelity. A threshold setting of zero results in a lossless compression strategy since no deviation from the average value is allowed and hence all the samples in the region must be identical. When using a threshold of zero, however, the compression ratio will typically be negative and thus of no practical use.

Figure 10.9 provides a numeric illustration. A continuous tone grayscale image is shown in (a) and encoded with a standard deviation threshold of 5.0. The standard deviation of the 4×4 image is computed as $\sigma = 31.3$, which exceeds the threshold of 5.0 and so the image is subdivided into four quadrants. The standard deviation of the four quadrants is then computed as $\sigma_{Q_0} = 2.3$, $\sigma_{Q_1} = 0.5$, $\sigma_{Q_2} = 4.6$, and $\sigma_{Q_3} = 31.34$. Since the first three quadrants have acceptable deviations, they are not further subdivided and are represented by their averages. The fourth quadrant, however, has an unacceptably large standard deviation and is hence subdivided into four 1×1 quadrants. In order to determine the fidelity of the compressed image we take the difference between the two images, as shown in Figure 10.9(c) from which the e_{rms} error is then easily computed as .649.

Figure 10.10 provides a visual example of this compression strategy. A continuous tone grayscale image is compressed with a threshold setting of 20 to obtain the image of (b). This threshold setting indicates that, on average, no sample of the compressed image will be more than 20 grayscale values different than the corresponding sample of the source image. A threshold of 10 is used to obtain the compressed image of (c) while (d) superimposes grids over regions of constant tone. The image of Figure 10.10 part (d) highlights the fact that CAC encoding dynamically allocates bandwidth (memory) to those areas of the image that contain high frequency information and less bandwidth to those areas of the image that do not posses a great deal of visual content.

(a) Source image.

(b) Threshold of 20.

(c) Threshold of 10.

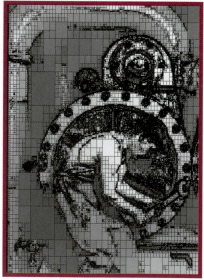

(d) Threshold of 10 with solid
areas outlined.

Figure 10.10. The effect of threshold on CAC compression.

10.4 Predictive Coding

10.4.1 Lossless Predictive Coding

Predictive coding reduces interpixel redundancy by recognizing that neighboring pixels are usually similar and that it is possible to predict with good accuracy what value a pixel will be if we know the preceding pixel values. Predictive encoding thus only encodes the difference between a sample's expected (or predicted) value and its actual value. Predictive coding schemes can be modeled, as shown in Figure 10.11, where a stream of samples is fed into the encoder. The predictor is responsible for predicting the next sample in the stream and is able to make good predictions since it has access to the previous samples in the encoding stream. The prediction may be real valued and hence rounding is used to generate an integer-valued prediction, which is subtracted from the actual sample value and the difference is then encoded by the symbol encoder.

An image decoder reads in the stream of differences, which are symbolically decoded and added to the predictions made by the predictor. The predictor has access to the stream of decoded samples in the same manner as the encoder predictor. Note that the predictor block of both the encoder and decoder must be identical for this scheme to work.

Consider a predictive encoding scheme for grayscale images. Given a stream of samples obtained through some scan of the image data, the predicted value

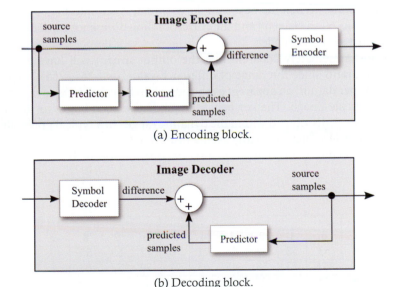

(a) Encoding block.

(b) Decoding block.

Figure 10.11. Block diagram of (a) predictive encoding and (b) predictive decoding.

of the kth sample is denoted as S'_k while the actual value of the kth sample is designated as S_k. The predictor block embedded in the encoding module must be able to compute S_k solely from the previous $k-1$ actual sample values (i.e., from the sequence $S_0, S_1, \ldots, S_{k-1}$) since predictions must be based on an analysis of actual sample values. While it may be imagined that the predictor contained in the encoder block is able to predict S_k by analysis of the entire sample stream, this is not allowed since the encoding predictor and decoding predictor must be identical and the corresponding predictor of the decoding block will only have access to already decoded samples.

One of the simplest predictor schemes is to use the most recent actual sample value S_k as the value of the next predicted sample S'_{k+1}. Given a stream of N samples $S_0, S_1, \ldots, S_{N-1}$ we define $S'_k = S_{k-1}$. When using this predictor, the first predicted value must be pulled out of thin air since there is no actual sample value preceding the first. A reasonable approach is to define S_0 as the center of the dynamic range, or 128 in the case of 8-bit samples.

Figure 10.12 provides an example of encoding a sample stream. The top row of the table contains the actual sample stream that is fed into the encoder. The predicted values (middle row) are forwarded from the most recent actual sample and the difference between actual and predicted is recorded to the output stream (bottom row). The sample stream will generally correspond to a raster scan of the image although alternate scanning schemes are possible. When the first sample, $S_0 = 53$, is fed into the predictive encoder block, the predicted value, $S'_0 = 128$, is subtracted from the actual value and this difference, $e = -75$, is encoded in the output stream. When the next sample, $S_1 = 58$, is fed into the encoder, the predicted value, $S'_1 = S_0 = 53$ and the corresponding difference $e = 58 - 53 = 5$ is given to the symbol encoder.

It is not immediately apparent that the output stream will have a positive compression ratio since the range of difference values is [-255, 255], which requires 9-bits per datum; an apparent expansion of an 8 bit-per-sample raw encoding of a grayscale image. The key benefit in predictive encoding is an alteration of the way data is distributed within the dynamic range. In a typical grayscale image, the grayscale values are nearly evenly distributed across the dynamic range. The effect of predictive encoding is to alter the distribution such that the sample

					K					
	0	**1**	**2**	**3**	**4**	**5**	**6**	**7**	**8**	**9**
S	53	58	45	63	71	70	72	68	55	...
S'	128	53	58	45	63	71	70	72	68	...
e	-75	5	-13	18	8	-1	2	-4	-13	...

Figure 10.12. Predictive encoding example.

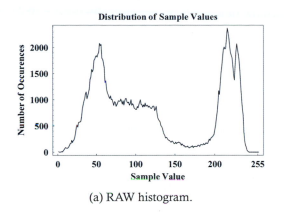
(a) RAW histogram.

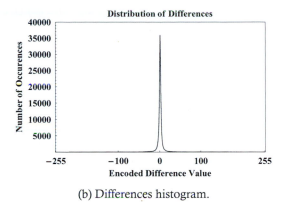
(b) Differences histogram.

Figure 10.13. Distribution of predictively encoded data.

values are highly clustered around zero. Figure 10.13 illustrates how the distribution of grayscale samples, shown in the histogram of (a) is altered to the zero-centered distribution of differences shown in (b). The predictor module will generally produce data that is tightly distributed about zero and falls off dramatically with distance from zero.

A specialized symbol encoding scheme can then be devised that is tuned for the expected distribution of data. One such scheme is to encode the data into a combination of 4-bit and 8-bit words. Since there are 16 patterns in a 4-bit word, we can represent any difference in the range $[-7, -6, \ldots, 5, 6]$ while leaving two extra patterns to designate whether the difference is negative or positive and then encode the magnitude of the difference as an 8-bit word. This coding scheme is given in Table 10.1.

When a difference of -5 is encoded, Table 10.1 outputs the 4-bit pattern for the decimal value 2 or 0010. When a difference of 128 is encoded we output the 4-bit pattern 1111, or decimal 15, followed by the 8-bit pattern of 10000000. While any difference outside of the $[-7, -6, \ldots, 5, 6]$ range will consume 12 bits of memory, the overwhelming majority of differences will fall within the

Difference Value	Coding Scheme
$-7 \ldots 6$	encode a 4-bit representation of the difference using 0 - 13
$-255 \ldots -8$	encode the 4-bit value 14 followed by the 8-bit difference
$7 \ldots 255$	encode the 4-bit value 15 followed by the 8-bit difference

Table 10.1. Table for encoding difference values.

near-zero range and hence consume only 4 bits of memory with the net effect of reducing the bits per sample.

10.4.2 Delta Modulation

Lossy predictive coding schemes reduce memory consumption by encoding the *approximate* differences between predicted and actual sample values. In the simplest such scheme, known as delta modulation, the error between a predicted and actual sample value is approximated as either $+\delta$ or $-\delta$ where δ is a global parameter established prior to compressing an image. While the predictor model of Figure 10.11 must be slightly modified in order to incorporate the notion of difference approximation, the process of delta modulation coding is easily described.

Consider an image with a sample stream of $\{53, 58, 45, 63, 71, 70, 72, 68, 55, \ldots\}$. Delta modulation will seed the encoder by passing the initial sample value, 53, directly to the output stream. Having recorded the 0th actual sample value, the predicted value of the $(k+1)^{\text{st}}$ sample, S'_{k+1}, is given as $S'_k + e'_k$ where e'_k is the approximate error between S'_k and S_k. The approximate error e'_k is defined as

$$
e'_k = \begin{cases}
0 & \text{if } k \leq 0, \\
+\delta & \text{if } S_{k+1} > S'_k \text{ and } k > 0, \\
-\delta & \text{otherwise.}
\end{cases}
$$

Figure 10.14 illustrates how an encoder would use delta modulation to represent the entire data stream of this example. The parameter δ is a global entity that can be recorded in the header of the output stream along with the initial sample value. Since the approximate differences are defined as either positive or negative δ, the only item that needs to be recorded in the output stream is the sign of the difference approximation. A single bit will suffice to distinguish between the symbols $+\delta$ and $-\delta$ with the net effect of reducing the bandwidth to one bit per sample.

The overall effect of this technique on the output is dependent on the value of δ. The value of δ determines the maximum amount that adjacent samples are allowed to change and, in fact, must change. Smaller delta values will therefore not be able to represent strong edges since samples vary quickly at edge

	0	1	2	3	4	5	6	7	8	9
S	53	58	45	63	71	70	72	68	55	...
S′	-	53	61.5	53	61.5	70	61.5	70	61.5	...
e′	0	$+\delta$	$-\delta$	$+\delta$	$+\delta$	$-\delta$	$+\delta$	$-\delta$	$-\delta$	...

Figure 10.14. Delta modulation example.

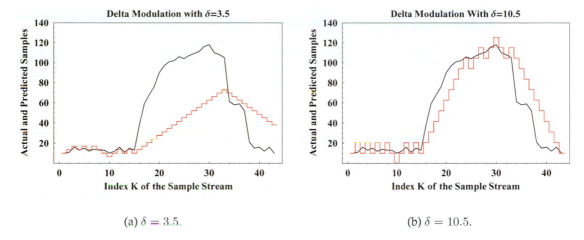

(a) $\delta = 3.5$.   (b) $\delta = 10.5$.

Figure 10.15. Graphs of delta modulation.

boundaries. Smaller delta values will, however, more accurately capture the fine details in a source image. Larger delta values will allow the compressed image to adequately represent strong edges, but will not allow the compressed image to adequately represent homogenous regions since the output will oscillate above and below the actual values of the region.

Figure 10.15 shows graphical representations of a sample stream that is encoded using delta modulation and two different delta values. The step-wise component in each graph represents the encoded sample stream while the smooth component represents a continuous tone grayscale edge going from near-black to mid-level gray and back to near-black. With the relatively low delta value of 3.5 in (a), the encoded representation can closely track the initial flat region of the input but is unable to track edges as the center of the graph shows. In part (b), however, the relatively high delta value of 10.5 allows the encoded data to more closely track the edge but sacrifices fidelity in the initial flat region of the input.

As described earlier, the initial sample value may be written in the header of the encoded stream. Far better results are obtained, however, by writing the first grayscale value of each row to the encoded stream. This allows the decoded image a greater coverage of grayscale values but does slightly reduce the compression ratio. Figure 10.16 shows a grayscale image that has been delta modulated using $\delta = 6.5$ in (b), $\delta = 20$ in (c), and $\delta = 75$ in (d). Note the blurring effect due to the relatively low delta setting in (b) while part (d) has sharper edges but appears grainy due to the large oscillations between adjacent samples.

(a) Grayscale source image.   (b) Compressed with $\delta = 6.5$.

(c) Compressed with $\delta = 20$.   (d) Compressed with $\delta = 75$.

Figure 10.16. Delta modulation example.

10.5   JPEG Case Study

JPEG compression is a sophisticated standard developed by the Joint Photographic Experts Group, a standards-making body comprised of industrial and academic partners. The JPEG standard describes a family of compression techniques that can be both lossless and lossy, although one particular implementation, JPEG/JFIF, is the most well-known realization of the standard and is a lossy technique. The JPEG standard was designed for use on photographic or natural scenes rather than synthetic graphic or iconic images. JPEG compression is the default format for most commercial digital cameras and is supported by most commercial photo editing software suites.

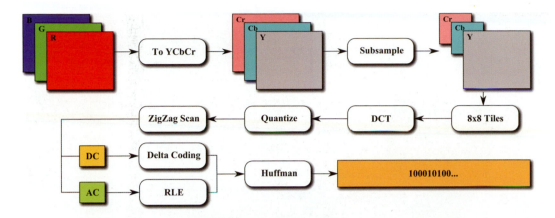

Figure 10.17. JPEG encoder.

JPEG compression applies a variety of techniques to maximize the compression ratio. Figure 10.17 gives a simplified block diagram of a JPEG encoder following the JFIF implementation where data elements are represented by colored rectangular boxes and processes are denoted by rounded boxes. The source image is first converted into the YCbCr color space after which the Cb and Cr bands are optionally subsampled. Each of the YCbCr bands is then tiled into 8×8 blocks such that if the source image dimensions are not multiples of eight then the image is padded with black borders and those borders ignored when decoded. Each 8×8 block is then converted into the frequency domain through the DCT transformation, which produces 64 DCT coefficients. The 64 DCT coefficients are then quantized by dividing each of the coefficients by values contained in a quantization table. Subsampling and quantization are the only portions of the JPEG/JFIF pipeline that are lossy and the degree of loss is relative to the quantization table itself.

Recall that the DCT transform tends to compact information into the upper-left values of the two-dimensional coefficients and hence the bottom-right coefficients are typically much smaller than those of the upper-right. In practice, most of the bottom-right coefficients are quantized to zero so that a zigzag scan of the AC coefficients of the 8×8 tile of quantized DCT coefficients typically produces long runs of zero values, which can be effectively run length encoded. The DC values of successive 8×8 blocks are typically similar and hence they are encoded using the predictive encoding scheme described in Section 10.4. Finally, the entire encoded stream is compressed further using huffman compression to mitigate any coding redundancy that remains.

The JPEG/JFIF standard works in the $YCbCr$ colorspace so that an RGB image is first converted prior to entering the encoding pipeline. The JPEG/ JFIF implementation also allows for less bandwidth to be allocated by the color

Figure 10.18. JPEG example where compression quality varies from 0 on the far left to 1 on the far right.

channels and hence the Cb and Cr channels are normally downsampled by a factor of two since human perception is not as sensitive to changes in color as it is to changes in light intensity. Most JPEG compressors allow the user to set the quality level. The level is generally a single value in the range $[0, 1]$ where 0 is expected to produce large compression ratios at the cost of quality while a setting of 1 gives low compression but high quality. Since quantization is the only part of the process that is lossy, the quality parameter directly controls the values in the quantization table. If all of the values in the quantization table are set to 1, there is no loss of information during quantization and the resulting JPEG compressed image will be relatively large but be a perfect representation of the source. Figure 10.18 shows the effects of JPEG compression. The image of this figure is a composition of JPEG compressed images as the quality setting is varied from 0, on the far left, to 1 on the far right.

The JPEG standard was updated by the standards committee in the year 2000 and is appropriately named JPEG 2000. The JPEG 2000 standard is based on a wavelet decomposition rather than a discrete cosine decomposition of source image. JPEG 2000 provides slightly better fidelity than the JPEG standard at

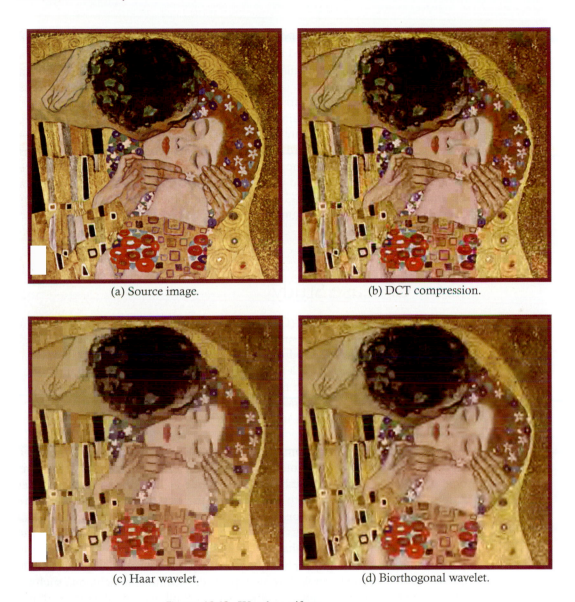

(a) Source image.   (b) DCT compression.

(c) Haar wavelet.   (d) Biorthogonal wavelet.

Figure 10.19. Wavelet artifacts.

low compression ratios but much better fidelity at high compression ratios. The 2000 standard is based on wavelets and does not subdivide the source into tiles. Progressive transmission is therefore naturally supported since coarse scalings of the original image are contained in the first bytes of the compressed data stream while the detailed coefficients follow. This allows decoders to dynamically generate increasingly more detailed results as the data becomes available.

The wavelet-based technique is nearly identical to the DCT technique in terms of the overarching idea. As described in Chapter 9, the DWT converts a sequence into low and high frequency coefficients. The high frequency coefficients can be truncated to obtain increasingly large compression ratios as more coefficients are dropped. Figure 10.19 shows a visual comparison between the JPEG and JPEG 2000 standard as a source image is compressed using various wavelets. In each case the compression ratio is approximately 96:1. The source image of part (a) is compressed using the DCT-based JPEG standard to obtain the image of part (b). The JPEG 2000 technique is then employed using a Haar basis in part (c) and a biorthogonal basis in part (d).

10.6   GIF Case Study

The *graphics interchange format* (GIF) uses an 8-bit indexed encoding scheme to represent image data and also supports animation by embedding several images in one file. CompuServe introduced the GIF format in 1987 and it quickly became popular because it used a form of data compression, LZW compression, which was more efficient than the run length encoding in common use at the time. GIF was one of the first two image formats commonly used on websites while JPEG support was introduced later with the release of the Mosaic browser.

GIF is generally a lossy compression technique that reduces the number of colors in an image to a palette of no more than 256 unique entries through some type of color dithering or thresholding technique as described in Section 8.7. After reducing a 24-bit image to 8 bits or less, the resulting data stream is compressed using LZW compression. For source images that have no more than 256 unique colors, GIF is a lossless compression scheme. GIF is therefore best used for iconic or synthetic images that have only a few total colors.

Encoding a GIF image simply requires that a palette of colors be given, which is then used in conjunction with some type of error diffusion technique such as Floyd-Steinberg. Some ad hoc standard palettes are often used, of which perhaps the most popular is commonly referred to as the "web-safe" color palette. This color scheme was developed at a time when most commercial displays were only able to support 256 colors. In the web-safe palette each of the red, green, and blue bands supports six shades of intensity for a total of $6 \cdot 6 \cdot 6 = 216$ total colors. The palette doesn't support 256 colors since many computing platforms reserved a handful of colors for custom use. The colors of this palette were chosen primarily since the popular web browsers of the time supported them.

The fidelity of a GIF-encoded image can be significantly improved, however, by using a palette that is carefully selected and optimized for the image that is being encoded. This presents an interesting preprocessing step in GIF encoding which is to compute the palette that will maximize the fidelity of the encoded image. In other words, given a source image I we must generate a list of no more than 256 distinct colors such that when I is dithered with those colors, it will produce the most visually pleasing result.

The median cut algorithm is a method for color quantization that works by finding clusters of similar colors and representing all colors in the cluster by a single color. The algorithm is straightforward to describe and is given in Listing 10.6. The inputs are an image and a value N that represents the desired size of the resulting palette. The algorithm works by viewing each color of the image as a point in the three-dimensional RGB color space. The central idea is to construct N boxes of approximately equal dimension each of which contains approximately the same number of colors, or source pixels. Associated with each box is a color that is given as the center of the box itself.

```
1  Palette algorithm(Image SRC, int N) {
2      let B be a set of boxes containing points in RGB space
3      let BB be the smallest box containing all colors in SRC
4      insert BB into B
5      while the number of boxes in B is less than N
6          let L be the largest length of any side of any box BC
7          cut BC into two boxes along its largest side
8              such that half of the contained points fall into each new box.
9          shrink the two new boxes so that they are just large enough
10             to contain their points
11     return the colors corresponding to the boxes of B
```

Listing 10.6. Median cut algorithm.

The method initially finds the bounding box of all colors in the source image. The box is then divided by cutting the longest side at the location of the median color in that dimension. Once the box has been split in half, each of the halves is then shrunk if necessary so that they are as small as possible while still containing all colors, or points, within their respective interiors. This process is repeated by finding the longest side of any box and dividing that box as described until there are exactly N boxes. The colors corresponding to these boxes then comprise the palette. Figure 10.20 illustrates the effects of dithering a source image (a) using a limited palette of four colors (b) and a larger eight-color palette in (c).

(a) Source image.   (b) Four color dithering.   (c) Eight color dithering.

Figure 10.20. Example of GIF dithering.

10.7   Digital Data Embedding

Steganography is a flourishing area of study within computer science and covers a broad range of specialties. In addition, the increasing use of digital images and image processing techniques makes the study of digital image processing an important subdiscipline of computer science. These two fields are brought together in a specialized application of steganography whereby secret messages are imperceptibly embedded in digital images. Steganography can be defined as "the art of concealing the existence of information within seemingly innocuous carriers." Steganography, in essence, camouflages a message to hide its existence and make it seem invisible thus concealing altogether the fact that a message is being sent. An encrypted message may draw suspicion while an invisible message will not.

The study of steganography has become increasingly important since the events of September 11, 2001. Shortly after the attacks on the World Trade Center towers in New York City, it was reported that Osama Bin Laden and his allies had communicated, in part, through the use of secret messages embedded in publicly distributed pornographic images. Researchers quickly sought to develop methods of detecting whether images contained hidden messages and how to decode them if they were present. While this provided the impetus for a sensationalist involvement in the study of steganography, more legitimate and profitable uses of steganographic techniques involve watermarks and digital copyright management. Digital watermarking is a technique for embedding information transparently into digital media. The embedded information is often a digital signature that uniquely identifies a copyright owner. Identifying information must be difficult to detect without possession of some key data, and it must uniquely identify the legitimate copyright owner, and must be difficult to erase or alter even if detected.

This section provides the technical background necessary for understanding basic steganographic techniques. The technique described here is the well-known least significant bit (LSB) approach. More sophisticated methods involve spread-spectrum encodings where a message is embedded across a frequency-based representation of a digital image. These spectrum-based approaches are beyond the scope of this text and will not be covered here.

10.7.1 Digital Image Representation

A brief review of binary representation will be instructive when interpreting bit-level pixel data in the context of data embedding. An 8-bit binary number has the general form

$$a_7 + a_6 + \ldots + a_1 + a_0,$$

where each a_n represents a single binary digit. In a digital image it is clear that a_7 is the most significant bit and indicates whether the pixel value is greater than 127 or not. A common means of converting a grayscale image to a binary image, for example, is to extract the a_7 bitplane. By contrast, a_0 embodies relatively little information and, in the context of a digital image, can generally be understood as a *noise* channel. The more significant bits of each pixel embody greater structural information than the least significant bits.

10.7.2 LSB Embedding

The human eye has a simultaneous resolution far below 8-bits which implies that much of the information contained in a digital image is imperceptible to a human observer and that small modifications to a source image will not be perceived. Steganography, in the simplest case, capitalizes on this overabundance of information by replacing the noise channels (i.e., the least significant bit channels) with data of the user's choosing.

Figure 10.21 gives an overview of one type of steganographic process flow. In this illustration an 8-bit source image, hereafter referred to as a cover, is viewed as eight separate channels. Data is directly copied from the cover image into the output image, also known as the stego image. As the data is copied, however, certain of the channels are blocked and replaced by the secret message. The secret message replaces the least significant channels (in this case the three least significant channels) of the source in order to produce the stego image. The

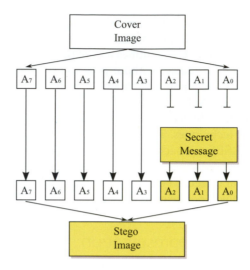

Figure 10.21. Embedding of a secret message into the three least significant channels of a cover image.

stego image will generally be visually indistinguishable from the cover image but is nonetheless different than the cover since the injected message has distorted the three least significant bit planes.

The degree to which the resulting stego image visually resembles the cover is dependent upon the number of channels that are used to encode the secret message. In practice, the fidelity of the stego image quickly degrades if more than three channels are used as secret message carriers. Also, while the content of the secret message itself affects the visual quality of the result, the effect is relatively less important than the number of channels used to inject the message. While beyond the scope of this article it is worth noting that adaptive techniques exist that allocate carrier channels on a per-pixel basis through evaluation of image texture or morphological structure. In other words, more bits are overwritten in pixels that may be greatly altered without perception and fewer bits are written in pixels where small alterations are more readily noticed.

Consider a 2 row by 8 column image with pixel values, as shown in Table 10.2. The image can be visualized as generally light-gray towards the left descending towards dark-gray towards the right of the image. The image is given in decimal notation in (a) and in binary notation in (b).

The secret message "HI" is then embedded into the cover. The message, when represented as ASCII text corresponds to the two-byte sequence of numerical values $\{72, 73\}$ or represented in 8-bit binary notation, $\{01001000, 01001001\}$. The secret message is actually viewed as a sequence of 16 individual bits that are injected into the noise channel(s) of the cover image. Since the cover image con-

| 209 | 200 | 188 | 192 | 138 | 131 | 99 | 50 |
| 198 | 155 | 140 | 187 | 154 | 130 | 122 | 100 |

(a) Decimal notation

| 11010001 | 11001000 | 10111100 | 11000000 | 10001010 | 10000011 | 01100011 | 00110010 |
| 11000110 | 10011011 | 10001100 | 10111011 | 10011010 | 10000010 | 01111010 | 01100100 |

(b) Binary notation.

Table 10.2. A 2 row by 8 column grayscale cover image.

tains 16 total samples the single least significant bit of each pixel is designated as a carrier of the secret message. In other words, the cover image dedicates a single channel towards carrying the secret message.

Table 10.3 gives the result of the embedding where the secret message has been distributed over the least significant bits of the cover. Note that many of the cover pixel values have not been altered at all in the process. Only eight of the cover samples have been modified by the embedding process and in those cases the samples have been modified by a single, imperceptible grayscale level.

While in this case the secret message was textual, any type of serializable data can be embedded into a digital image using this technique. A practical steganographic tool would include header information prior to the secret message itself all of which has been encrypted and compressed in a preprocessing phase.

| 11010000 | 11001001 | 10111100 | 11000000 | 10001011 | 10000010 | 01100010 | 00110010 |
| 11000110 | 10011011 | 10001100 | 10111010 | 10011011 | 10000010 | 01111010 | 01100101 |

Table 10.3. Stego image with embedded ASCII "HI."

10.7.3   Implementation

Fundamental to digital data embedding are the abilities to encode a secret message into a cover and to extract a secret message from a stego image. The Steganographer interface shown in Figure 10.22 captures these responsibilities in the embed and extract methods. The embed method accepts a BufferedImage which serves as the cover and an Object serving as the secret message. The method creates a stego image, embedding the secret message into the cover and returning the result as an BufferedImage while leaving the cover image intact. The extract method serves as an inverse to the embed method, returning an Object that has been extracted from a stego BufferedImage. This interface serves as a generic and extensible solution since there is no functional dependency upon the way in which the embedding takes place. These methods simply

Figure 10.22. UML class diagram of the Steganographer interface.

serve as hooks for customized steganographic algorithms that are provided by concrete Steganographer implementations.

In order to simplify implementation a restriction not defined at the Steganographer level is imposed by the embed method. In particular, the message must be a BufferedImage rather than, in the most general sense, a serializable object. While the more general implementation is obviously preferable, it is not feasible in the context of an undergraduate computer science course. The embed method is allowed to throw an exception if the type of message specified is incompatible with this requirement.

Implementation of the Steganographer can be difficult in light of the primitive bit-level nature of the write and read operations. Implementation is greatly simplified, however, by clever design and the creation of an appropriate set of supporting classes. Towards this end two important classes, BufferedImage InputStream and BufferedImageOutputStream, are designed to isolate the low-level complexities and to supply a convenient interface to the necessary bitwise operations.

Figure 10.23 gives a UML class diagram of the BufferedImageOutput Stream. The BufferedImageOutputStream class provides a convenient way for clients to write data into a BufferedImage object via calls to the write method. The BufferedImage object that receives the data serves as a kind of memory output device rather than a file. When a client writes a sequence of bytes to the BufferedImageOutputStream, the data is written bit by bit into the samples of the destination in a raster-scan fashion starting from the least significant bit and moving onto the more significant bits whenever the need arises. The BufferedImageInputStream supplies the inverse operation, which reads data bit-by-bit from a BufferedImage, packages the data into bytes, and supplies the data to clients on demand.

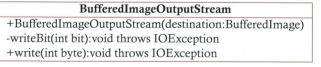

Figure 10.23. UML class diagram of the BufferedImageOutputStream.

Implementation of these two classes involves keeping track of where the next bit should be read from or written to and providing methods to read or write data on a bit-by-bit basis in the raster of a BufferedImage. Listing 10.7 gives a complete listing of an implementation of the output stream class. It is important to note that this class is a direct subclass of OutputStream and serves to override the write method by writing data into the destination image. The writeBit method serves as the fundamental method on which the write method is constructed. Although the writeBit method takes an integer as input, it treats only the least significant bit as meaningful. The LSB of the input integer is written to the destination image at the location given by indices x, y, *plane*, and *band*. After writing an individual bit these four indices are then updated to indicate the next bit location at which to write. If a writeBit method is called but the indices are invalid due to overflow then an IOException is thrown. This can only occur if a client writes more data into a destination image than the destination has capacity to store. The write method accepts an integer where the least significant byte is written to the output stream. In this implementation, the write method calls the writeBit method once for each of the eight bits that must be written. The BufferedImageInputStream class is not specified here but provides functionality inversely analogous to that of the BufferedImageOutputStream.

```
1  public class BufferedImageOutputStream extends OutputStream {
2      private BufferedImage destination;
3      private int x, y, plane, band, numberOfBands;
4      private WritableRaster raster;
5
6      public BufferedImageOutputStream(BufferedImage dest) {
7          destination = dest;
8          x = y = plane = band = 0;
9          raster=destination.getRaster();
10         numberOfBands=raster.getSampleModel().getNumBands();
11     }
12
13     public void write(int b) throws IOException {
14         for(int i=0; i<8; i++){
15             writeBit((b >> i) & 1);
16         }
17     }
18
19     private int setOn(int value) {
20         return value | (0x01 << plane);
21     }
22
23     private int setOff(int value) {
24         return value & (~(0x01 << plane));
25     }
26
27     private int setBit(int destinationSample, int bit) {
```

```
28          return bit==0?setOff(destinationSample):setOn(destinationSample);
29      }
30
31      private void advanceIndices() {
32        x++;
33        if(x==destination.getWidth()){
34          x=0;
35          y++;
36          if(y==destination.getHeight()){
37              y=0;
38              band++;
39              if(band==numberOfBands) {
40                  band=0;
41                  plane++;
42              }
43          }
44        }
45      }
46
47      private void writeBit(int b) throws IOException {
48        if(bitPlane > 7) throw new IOException("capacity exceeded");
49        b = b & 1;
50        try {
51          int sampleToChange = raster.getSample(x,y,band);
52          raster.setSample(x,y,band,setBit(sampleToChange,b));
53          advanceIndices();
54        } catch(Exception e) {
55          throw new IOException("error writing bit.");
56        }
57      }
58 }
```

Listing 10.7. Complete implementation of the `BufferedImageOutputStream`.

Once the input and output stream classes have been authored an implementation of the `Steganographer` interface is straightforward. We author a class named LSBSteganographer that performs data embedding in the LSB manner previously described by spreading the secret message across the least significant bits of the cover.

One of the key insights of this design is that the `ImageIO.write` method can be used to output an image to **any** output stream. Since the `BufferedImage OutputStream` is itself an output stream we can use the `ImageIO` method as an exceptionally convenient way of writing data into a destination image. Since the difficult work of storing an image is provided by the `ImageIO` class, the embed method of the LSBSteganographer is trivial to write and is completely implemented by the several lines of code given in Listing 10.8. This implementation will throw an `IOException` if the carrying capacity of the cover is not sufficient for embedding of the secret message.

```
1  public class LSBSteganographer {
2      public BufferedImage embed(BufferedImage cover, Object msg)
3              throws IOException {
4          BufferedImage result = cover.clone();
5          if(msg instanceof BufferedImage) {
6              BufferedImage secret = (BufferedImage)msg;
7              OutputStream output = new BufferedImageOutputStream(result);
8              ImageIO.write(secret, "PNG", output);
9          }
10         return result;
11     }
12
13 }
```

Listing 10.8. Implementation of the LSBSteganographer embed method.

It is important to note that the LSB embedding technique is effective only when using lossless file formats such as PNG. If the stego image is compressed using the JPEG/JFIF or GIF file formats, the secret message will be lost. This gives rise to an interesting variant of LSB encoding when using JPEG compression. It is possible to apply the LSB technique directly to the quantized DCT coefficients rather than the spatial domain samples. Since the quantized DCT coefficients are losslessly encoded, the secret message will not be mangled as it passes through the remainder of the JPEG encoding pipeline as depicted in Figure 10.17. In addition, the higher frequency coefficients can be altered with less visible effect than the low frequency coefficients and although the compression ratio of the resulting image will not be as high as if the message had not been injected into the source, the secret message is fully recoverable.

10.8   Exercises

1. Pixel Jelly supports a variety of image file formats as described in this text. Use Pixel Jelly to save full color, grayscale, and binary images using run length encoding. Analyze the results in terms of the degree of compression with respect to the RAW format.

2. Pixel Jelly supports a variety of image file formats as described in this text. Use Pixel Jelly to save full color, grayscale, and binary images using JPEG encoding varying the quality setting from 0 to 1 in .1 increments. Analyze the results in terms of compression achieved and determine the e_{rms} as a measure of the fidelity of the compressed image.

3. Pixel Jelly supports a variety of image file formats as described in this text. Use Pixel Jelly to save full color and grayscale images using lossy CAC encoding while varying the standard deviation from 0 to 20 in four unit increments. Analyze the results in terms of compression achieved and fidelity of the compressed image.

4. Write a program to run length encode/decode a BufferedImage. You should use the technique described in this chapter to support binary, grayscale, and true-color images. You should run your compression routine on a broad sampling of test images and record the compression ratios obtained where the compression ratios are defined with respect to the raw memory consumption.

5. Write a program that uses lossless hierarchical compression to encode/decode a BufferedImage. The program should only support binary images. You should execute your program on a suite of test images and record the compression ratios achieved.

6. Write a program that encodes/decodes a BufferedImage using lossy hierarchical compression. You should execute your program on a suite of test images and record both the compression ratio achieved and the rms error as a function of the quality threshold setting.

7. Write a program that encodes/decodes a BufferedImage using delta modulation such that the delta value can be set by the user. You should execute your program on a suite of test images and record both the compression ratio achieved and the overall rms error as a function of the delta value.

8. Given the encoding scheme of Table 10.1, determine what percentage of the errors must be within the range $[-7, 6]$ in order to achieve a positive compression ratio for predictive encoding. What is the maximum compression ratio that can be achieved using this technique and what is the lowest ratio that will be achieved? Provide robust proofs for your results.

Artwork

Figure 10.2. *Symphony in White no. 1 (The White Girl)* by James Abbott McNeill Whistler (1834–1903). James Whistler was an American painter whose work aroused significant controversy and whose work is believed to have paved the way for later art abstraction. Whistler was born in Lowell, Massachusetts, but moved to St. Petersburg, Russia as a youth. It is noteworthy that the "The White Girl" of Figure 10.2 was the subject of controversy not due to its objective nature but because of the subject. Gustave Courbet completed a pornographic work entitled *L'Origine du monde* in 1866 and the model he used for that graphic image is believed to have been Joanna Hiffernan, who is also the model for Whistler's "White Girl."

Figures 10.5 and 10.6. "Orson Welles" by Carl Van Vechten (1880–1964). See page 120 for a brief biographical sketch of the artist. The Welles portrait is available from the Library of Congress Prints and Photographs Division under the digital ID cph 3c19765.

Figure 10.10. *Power House Mechanic Working on Steam Pump* by Lewis Hine (1874–1940). See page 50 for a brief biography of the artist. The mechanic of Figure 10.10 is certainly the most iconic of Hine's work. This image is available from the National Archives and Records Administration, Records of the Work Projects Administration under ID 69-RH-4L-2.

Figure 10.16. *In Glacier National Park* by Ansel Easton Adams (1902–1984). Ansel Adams was an American photographer of the American West. His black-and-white photographs of natural settings, especially those of Yosemite National Park and the Grand Tetons, propelled his popularity. He was an active environmentalist and also something of a social activist. During the Japanese internment following the attack on Pearl Harbor, Adams put together a photo essay, subsequently exhibited in the Museum of Modern Art, that promoted the plight of Japanese Americans. The image shown in Figure 10.16 was commissioned by the National Park Service and is available from the National Archives using ID 79-AAE-23.

Figure 10.18. *Still Life* by Willem Claeszon Heda (1594–1680). Willem Heda was a Dutch artist devoted exclusively to still life painting. His style was precise and his compositions showed considerable attention to detail. He focused on extravagant place settings of oysters, salmon (as shown in Figure 10.18), lemon, bread, and champagne. It is unfortunate but pedagogically necessary that the image shown here is shown in such distorted form.

Figure 10.19. *The Kiss* by Gustav Klimt (1862–1918). See page 253 for a brief description of the artist and this work.

Figure 10.20. *Saint Catherine of Alexandria* by Raphael Sanzio (1483–1520). Raphael Sanzio was an Italian painter and architect during the High Renaissance. Sanzio, da Vinci, and Michelangelo, constituted the three great masters of that period. Sanzio was exceptionally prolific and left behind a large body of work despite his death at the early age of 37. Sanzio traveled a great deal during his career, working in various regions of Northern Italy and Florence. He was extremely influential in his time but his influence yielded to the immense popularity of Michelangelo.

Morphological Image Processing 11

The term *morphology* can be used in linguistics to refer to the study of the structure and form of words, in biology to the study of the form or shape of an organism, in astronomy to the shape of astronomical object or in geography to the study of the shape and structure of land masses and rivers. In the context of image processing, mathematical morphology treats image components as geometrical shapes. Mathematical morphology is concerned with how to identify, represent and process the shapes of objects within a scene. Morphology relies on several branches of mathematics such as discrete geometry, topology, and differential equations and has produced sophisticated and efficient algorithms for such tasks as handwriting analysis, biomedical imaging, and others. The material presented in this chapter begins to move away from the domain of image processing and into the more abstract domain of computer vision.

Recall that computer vision seeks to understand the semantic content of an image while image processing seeks to improve the quality of an image with respect to some problem-specific objective. In this context, image processing techniques are used to preprocess an image for higher level analysis. Segmentation is a process that identifies the objects within an image and classification describes the content of an image.

Figure 11.1 provides an example. The binary image in this figure has been acquired, binarized, and preprocessed through image processing techniques. Segmentation tells us that there are two clearly delineated shapes in the image which we have labeled as A and B. Shapes within an image are often referred to informally as *blobs* or, more formally, as *components*. Classification then tells us that component A is a square centered near the top while B is an elongated rectangular shape near the center.

The fundamental goals of morphological processing are those of preprocessing, segmenting, and classifying images and image components. In more specific terms, we seek to identify the components within an image, to find the boundary of a shape, to grow or shrink shapes, to separate or join shapes, to represent

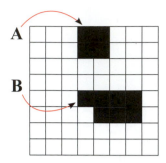

Figure 11.1. Image components.

shapes in some compact form, and to describe the essential properties of a shape. In this chapter we will emphasize these morphological techniques from a computational perspective rather than the set-theoretic approach of mathematical morphology. We will also introduce the morphological operations by limiting our attention to binary images and only briefly discuss how these techniques can be generalized for grayscale and color images near the end of the chapter.

11.1   Components

A component is a set of pixels where membership in the set is determined by both the color and the location of the pixel. A binary image contains only two colors, which are here classified as the foreground and the background. Any pixel that is a foreground pixel is a member of some component while no background pixel is part of a component. Throughout this text we assume that the foreground is black while the background is white. We also note that foreground pixels that are spatially connected belong to the same component while foreground pixels that are not spatially connected are not part of the same component. This spatial connectedness is known as *connectivity*.

Consider two adjacent pixels within a binary image. A pixel p at location (x, y) has neighboring pixels to the north, south, east, and west that are given by locations $(x, y - 1)$, $(x, y + 1)$, $(x + 1, y)$, and $(x - 1, y)$, respectively. These neighbors form a set of four pixels known as the *4-neighbors of p*, that we denote as $N_4(p)$. A pixel p also has four neighbors that adjoin to the northwest, northeast, southwest and southeast. These neighbors constitute a set of four pixels which we denote as $N_d(p)$ where the subscript d stands for the diagonal neighbors of p. The set of all eight of these neighbors is denoted as $N_8(p)$ and is referred to was the *8-neighbors* of p. The 8-neighbor set is the union of the 4-neighbor and diagonal neighbor sets: $N_8(p) = N_4(p) \cup N_d(p)$. Figure 11.2 illustrates each of these sets.

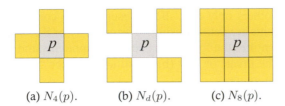

(a) $N_4(p)$.   (b) $N_d(p)$.   (c) $N_8(p)$.

Figure 11.2. Neighbors.

A pixel p is said to be 4-adjacent to another pixel q if q is one of the 4-neighbors of p. In other words, p is 4-adjacent to q if $q \in N_4(p)$. Similarly, a pixel p is said to be 8-adjacent to pixel q if q is one of the 8-neighbors of p. A pixel p is said to be 4-connected to pixel q if p is 4-adjacent to q and both p and q are the same color. Of course, p is 8-connected to q if p is 8-adjacent to q and they are the same color. Since we are now concerned with only binary images, pixel p is connected to pixel q only when both p and q are foreground pixels.

Pixels p and q are also said to be connected if there is a path of connected pixels extending from p to q. A path is a sequence of pixels $p_1, p_2, \ldots, p_{n-1}, p_n$ such that each pixel in the sequence is connected to the preceding (except for the first pixel since there is no preceding pixel) and succeeding pixel (except for the last pixel since there is no succeeding pixel) in the sequence. Connectedness, as used in the definition of path, refers to either 4-connectedness or 8-connectedness but not to both within the same path. We can then say that p is 4-connected to q if there exists a 4-connected path between p and q. We can also say that p is 8-connected to q if there exists an 8-connected path between p and q. We then formally define a *4-connected component* containing a pixel p as the set of all pixels that are 4-connected to p. An *8-connected component* containing pixel p is then naturally defined as the set of all pixels that are 8-connected to p.

Figure 11.3 illustrates how connectedness is used to identify components within a binary image. Given the binary source image of part (a) we seek to identify the components or shapes in the image. Some observers might identify

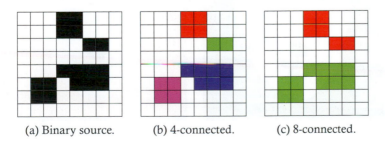

(a) Binary source.   (b) 4-connected.   (c) 8-connected.

Figure 11.3. Connected components.

four shapes and others only two. Both observers are correct depending upon whether components are defined using 4-connectivity or 8-connectivity. The image of part (b) identifies the four 4-connected components and labels them with unique colors. The image or part (b) identifies and labels the two 8-connected components in the image.

11.2   Component Labeling

One of the fundamental goals of morphological image processing is to automatically identify each component in an image. This task is known as component labeling. A label is any symbol that uniquely identifies a component and while integer labels are typically used, we will also use colors as labels in order to formulate the problem as an image processing filter. The problem of component labeling is presented as an image filter such that given a binary source image I we seek to generate a true color destination image such that each component of I is uniquely colored, or labeled. The central idea is shown in Figure 11.4, where the binary source image of part (a) is labeled in part (b). In this example, there are five separate components, each of which is uniquely labeled with a color. The largest component is labeled with red, the nose with blue, the lip consists of two separate components that are labeled with dark and light green and the small blob under the subject's right eye is labeled with gray. In each case, every pixel of the component is colored, or labeled, with the same color.

We begin by presenting a solution for a related problem known as flood filling. We describe flood filling as *given a foreground pixel p contained by some component in a binary source image, color all pixels of that component with a color C.*

(a) Source image.

(b) Labeled components.

Figure 11.4. Connected component labeling.

```
1  algorithm floodFill(Image SRC, Image DEST, int X, int Y, color C) {
2     if SRC(X,Y) is BACKGROUND return
3     if DEST(X,Y) is NOT BACKGROUND return
4
5     set the DEST sample at (X,Y) to C
6     floodFill(SRC, DEST, x-1, y, C)
7     floodFill(SRC, DEST, x+1, y, C)
8     floodFill(SRC, DEST, x, y-1, C)
9     floodFill(SRC, DEST, x, y+1, C)
10  }
```

Listing 11.1. Flood filling algorithm.

Flood filling can be concisely described with recursion, as shown in the algorithm of Listing 11.1. The floodFill method accepts a binary source image and a color destination image where every pixel is assumed to be initialized to the background color. The coordinates of pixel p are given by variables X and Y while the label, or color, is given as C, which is any color other than the background color of the source. The method finds every pixel belonging to the same 4-connected component as p and assigns it the color C in the destination.

Consider the logic underlying the flood fill algorithm. If location (X,Y) denotes a background pixel of the source image, then this pixel is not part of any component and nothing should be done to the corresponding destination pixel. Also, if location (X,Y) denotes a pixel that has already been colored in the destination then no action should be taken. These two cases constitute the base cases of the recursive algorithm and are reflected in lines 2 and 3. In all other cases, however, we should color the corresponding pixel of the destination with color C and proceed to flood-fill each of the 4-adjacent neighbors, in the case of 4-connected labeling, or the 8-adjacent neighbors, in the case of 8-connected labeling.

An algorithm for component labeling can then be described in terms of flood filling, as shown in Listing 11.2. This algorithm scans the source image and performs a flood fill on every foreground pixel that has not already been filled. Each label is a unique color excluding the background color.

The algorithm of Listing 11.2 unfortunately suffers from two important limitations. The first is a limitation on the efficiency or speed of execution. Since each pixel of the source may be probed multiple times the algorithm is expected to be approximately linear in the size of the image but with a possibly large constant multiplier. The second is a limitation on the reliability of the algorithm. Since some image components will contain millions of pixels, the recursive implementation is likely to overflow the execution stack of the processing system as it traverses deep paths through the component being labeled.

```
1   Image componentLabel(Image SRC) {
2     Image DEST = new color image
3
4     for every location (X,Y) in SRC
5       if SRC(X,Y) is a foreground pixel and DEST(X,Y) is a background pixel
6         LABEL = new unique random color that is not the background
7         floodFill(SRC, DEST, X, Y, LABEL)
8
9     return DEST
10  }
```

Listing 11.2. Connected component labeling algorithm

The union-find algorithm addresses both of these limitations by using a clever data structure that provides efficient support for the operations necessary when labeling components. The disjoint-set data structure is a structure that represents a collection of non-overlapping sets. In component labeling, the disjoint-set data structure stores a collection of components where each foreground pixel is a member of one of the components and no foreground pixel is a member of more than one component. The main virtue of a disjoint-set data structure is the support that it provides for the *find* and *union* operations. *Find* identifies the set that contains a particular element while *union*, as the name implies, merges two sets into a single set. In our presentation, the disjoint-set is represented as a forest of trees such that each node corresponds to a pixel, each tree represents a set of pixels, and the label of each tree is given by the root node. The find operation is implemented by finding the root node of a given pixel and the union operation corresponds to attaching the root node of one of the trees as the child of the other root.

The find and union operations are sufficient for implementing an efficient component labeling algorithm. A disjoint-set data structure is created and initialized by placing each foreground pixel into its own set. After this initial step, the source image is raster scanned and sets are unioned as necessary in order to form a collection of connected components. Consider the binary image of Figure 11.5, where the background color is white and the foreground color is black. There is then a single 8-connected component that should be identified at the conclusion of the component labeling algorithm. We will first describe the component labeling technique with abstract illustrations after which we will give a concrete implementation of this procedure in Java.

We first identify each pixel of Figure 11.5 according to its linearized index using a row-major layout scheme. There are then seven foreground pixels and each of them forms the root of a tree of size 1. The forest thus contains seven

Figure 11.5. Binary image for union-find example.

trees, or sets, having the labels 2, 4, 7, 8, 11, 13, and 14. We then perform a raster scan of the source image and for each foreground pixel p we examine all other pixels $q \in N_8(p)$ that have already been scanned. Excluding consideration of boundaries, there are exactly four 8-connected neighbors of p that have already been scanned: the pixels located to the northwest, north, northeast, and west of p. For every one of these pixels that is foreground pixel we find the set containing p and merge it with the set containing q since the pixels are 8-connected.

Figure 11.6 illustrates this process. Step A of this diagram shows the initial configuration of the forest such that there are seven disjoint sets where each set is a tree of size 1. The first foreground pixel that is connected to a pixel already scanned is pixel 7, which is merged with the set containing 2 to form the forest shown in step B. In each step of Figure 11.6 a red node denotes the label of a set, a green node denotes the pixel p that is being scanned, and a blue node denotes a set member that is not a label. The next foreground pixel that is connected to an already scanned pixel is 8 which is merged with 4 to obtain the forest shown

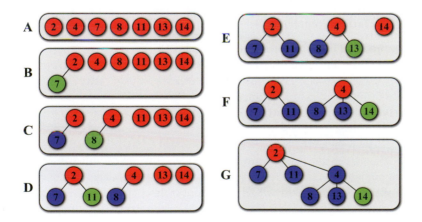

Figure 11.6. An example of the Union-find procedure.

in step C. This process continues until the two sets given as 2 and 4 in step F are both merged together by pixel 14, as shown in steps F and G. Note that when pixel 14 is scanned our illustration shows set 14 merged with set 4 (the set containing 13) to form the forest of step F. Pixel 14, however, is also connected to pixel 11 and so the final step is to merge set 4 (the set continuing 14) with set 2 (the set continuing 11) to form the forest of step G. Since at the conclusion of this algorithm the forest contains a single tree this indicates that there is a single 8-connected component having the label 2.

A disjoint-set can then be implemented as an integer array P such that pixels are denoted by their linearized index i and $P[i]$ is the parent node of pixel i. If $P[i] = -1$ then pixel i is background pixel. If $P[i] = i$ then pixel i is the root node of some tree. Of course, if $P[i] \neq i$ then pixel i is a child node of $P[i]$. For any background pixel i, the set containing that pixel can be found by following a path of parent references until reaching an index j where $P[j] = j$.

Consider Listing 11.3, which gives a complete implementation of the find and union operations within a class named `ConnectedCompents`. The find method accepts an integer index denoting a pixel and searches for the root node by recursively moving upward along the path given by the values in the parent array. Note that the find operation is linearly dependent on the length of the path from a pixel to the root. Since the find method visits each node as it climbs the path we can modify the tree by attaching each node along the path directly to the

```
1  public class ConnectedComponents {
2    private int[] parent;
3
4    private int find(int index) {
5      if(parent[index] == -1 || parent[index] == index) {
6        return parent[index];
7      } else {
8        parent[index] = find(parent[index]);
9        return parent[index];
10     }
11   }
12
13   private void union(int p1, int p2) {
14     int root1 = find(p1);
15     int root2 = find(p2);
16     parent[root2] = root1;
17   }
18
19   // other methods not shown
20 }
```

Listing 11.3. Find and union methods.

root. This modification is known as path compression since the tree is flattened as it is traversed. Path compression is implemented in line 8 of Listing 11.3 and is vital to efficient operation of the disjoint-set structure.

The union method is also given in Listing 11.3. This operations first finds the label of the set containing pixel p1 after which it finds the label of the set containing pixel p2. The set containing pixel p2 is attached as a child to the root node of the set containing p1 thus merging the two sets into one. The method is extremely efficient assuming that finds are efficiently encoded.

Figure 11.7 illustrates how this data structure is used to label the components in the image of Figure 11.5. Each step of the process is labeled from A to G in correspondence with the steps illustrated in the abstract illustrations of Figure 11.6. Figure 11.7 shows the contents of the parent array at the conclusion of each step and note that the parent array is how the forest of Figure 11.6 is encoded in our implementation.

In step A the parent array has been initialized such that for every background pixel parent[i] = -1 and that for every foreground pixel parent[i] = i. Just as before, the first pixel that must be unioned with another is pixel 7, which is merged with 2, as shown in step B. Pixel 8 is unioned with 2 in step C and the process continues exactly as described in Figure 11.6 until all pixels have been scanned and merged when necessary. The parent array shown in step F–G is the final configuration of component labeling where note that for any pixel i the label of that pixel is given by find(i).

In summary, the union-find algorithm in conjunction with the disjoint-set data structure provides near linear-time support for component labeling. Also, since the algorithm works on a raster-scan basis, the algorithm can be modified to label images whose resolution is so large that they cannot be made memory resident.

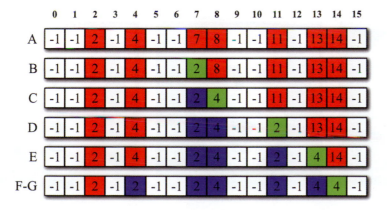

Figure 11.7. Union-find procedure with arrays.

11.3   Dilation and Erosion

Dilation and erosion are two fundamental image processing operations with respect to morphological processing. Dilation and erosion are low level operations upon which many more sophisticated operations rely. Dilation, as the name implies, causes a component to grow larger while erosion shrinks or thins a component. Figure 11.8 shows an example of these two operations. The binary source image of part (a) is dilated to obtain the result of (b) while part (c) shows the result of erosion.

The specific effect of each of these operations is controlled by a *structuring element* in much the same way that a kernel controls the specific effect of convolution. A structuring element is itself a binary image that is used to carve out the edges of a component, in the case of erosion, or to expand the edges of a component, as in dilation. Figure 11.9 shows a structuring element in part (a). The illustration here uses a non-binary representation for the structuring element but only for the sake of visual clarity. The structuring element in this example should be understood as a 3×3 binary image. The source image of (b) is then eroded with the structuring element to obtain the result shown in part (c).

When either dilating or eroding an image, the center of the structuring element is positioned above every source pixel in a raster scan of the source image. The output produced by the structuring element for the pixel over which it is positioned is then determined as either black or white. For erosion, the center pixel is black only when every black pixel of the structuring element is positioned over a black element of the source while for dilation the output is black if at least one black pixel of the structuring element is positioned over a black pixel of the source.

Figure 11.10 illustrates a number of cases involving the overlap of the structuring element with the source image. In part (a), the structuring element does

(a) Source image.   (b) Dilated.   (c) Eroded.

Figure 11.8. Dilation and erosion.

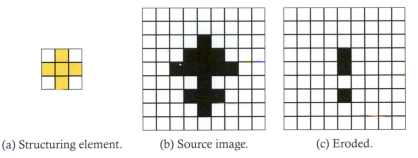

(a) Structuring element. (b) Source image. (c) Eroded.

Figure 11.9. Structuring element and erosion.

not overlap the source image at all and hence the output for both erosion and dilation is white. In part (b) the structuring element partially overlaps the source image and hence the output for erosion is white and for dilation is black. In part (c), the structuring element completely overlaps with the source image and hence the output for both erosion and dilation is black.

Consider how a wood router can be used to trim the edges of a piece of woodwork. The craftsman selects a routing bit of the correct size and shape and then sweeps the router along the boundaries of the wooden piece in order to thin, or erode, the wood. Erosion can be understood in this way where the structuring element corresponds to the routing bit and the process of sweeping along the edges of the wooden piece corresponds to the erosion operation itself. The most common structuring elements are either square or circular in shape where the circular elements tend to produce smoother edges than others as would naturally be expected.

Dilation and erosion are generally used as a preprocessing phase in shape analysis. Consider a binary image that has numerous 4-connected components. Erosion may increase the number of components by splitting a component into

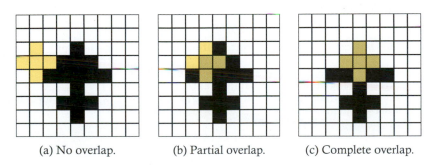

(a) No overlap. (b) Partial overlap. (c) Complete overlap.

Figure 11.10. Cases of overlap with the structuring element.

two or more separate shapes. Figure 11.9 illustrates precisely this effect where the source image of (a) contains a single 4-connected component prior to erosion but contains two 4-connected components after erosion. Dilation may, of course, have just the reverse effect of joining two source components into a single shape.

11.3.1   Implementation

Erosion and dilation can be implemented via use of the Mask class introduced for use in rank filtering. A mask is simply a binary image, implemented as a boolean-valued matrix that can also be scanned. Listing 11.4 is a complete implementation where the filter method simply iterates over every source sample and erodes that element by centering the structuring element over the pixel and scanning the neighborhood. Erosion of a single source sample is done via the erode method.

```java
public class ErosionOp extends NullOp {
  private Mask structuringElement;
  private ImagePadder sampler = new ColorFramePadder(Color.white);

  public ErosionOp() {
    this(MaskFactory.getCircleMask(7).invert();
  }

  public ErosionOp(Mask element) {
    setElement(element);
  }

  public Mask getElement() {
    return structuringElement;
  }

  public void setElement(Mask element) {
    structuringElement = element;
  }

  private void erode(BufferedImage src, BufferedImage dest, Location pt) {
  int result = 0;
  for(Location kpt : new RasterScanner(structuringElement.getBounds())) {
    int element = structuringElement.getSample(kpt.col, kpt.row);
    int sample = sampler.getSample(src, pt.col + kpt.col, pt.row + kpt.row, 0);
    if(sample != 0 && element == 0) {
      result = 255;
      break;
    }
  }
  }
```

```
31        dest.getRaster().setSample(pt.col, pt.row, 0, result);
32      }
33
34      public BufferedImage filter(BufferedImage src, BufferedImage dest) {
35        if(src.getType() != BufferedImage.TYPE_BYTE_BINARY)
36          throw new IllegalArgumentException("Source must be a binary image");
37
38        if(dest == null) {
39          dest = createCompatibleDestImage(src, src.getColorModel());
40        }
41
42        for(Location pt : new RasterScanner(src, false)) {
43          erode(src, dest, pt);
44        }
45        return dest;
46      }
47 }
```

Listing 11.4. ErosionOp.

11.4   Opening and Closing

Opening and closing are morphological processes based on dilation and erosion. Opening is used to separate a single component within a source image at narrow pathways between portions of the component. Closing has the opposite effect of joining two separate shapes that are in sufficiently close proximity. Opening and closing tend to otherwise leave the thickness of components unchanged, which makes these operations significantly different than either dilation or erosion.

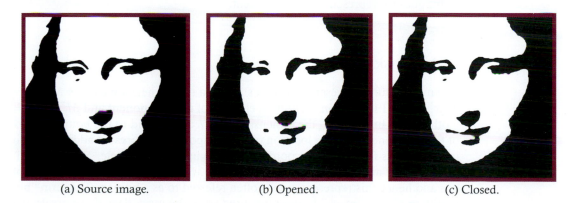

(a) Source image.   (b) Opened.   (c) Closed.

Figure 11.11. Example of opening and closing.

The effects of opening and closing are illustrated in Figure 11.11. The source image of part (a) is opened in part (b) and closed in part (c) using a circular structuring element of appropriate size. Notice in part (b) that the subject's right eyebrow has divided into two separate components and that the separation between the two lip components has increased. Notice in part (c) that the subject's nose has joined together with the two lip components to form a single shape. In general terms, however, note that neither opening nor closing has resulted in a significant enlarging or shrinking of the source component boundaries except along narrow paths in the source.

Opening an image A by a structuring element B is simply the erosion of A with B followed by the dilation of A with B. Closing is the dilation of A with B followed by the erosion of A with B.

11.5   Component Representation

While erosion and dilation affect the components within an image they are not applied to components as such but to binary images. Other operations are, however, specific to an individual component rather than a binary image. Operations such as translating a component, smoothing a component, determining the area or perimeter of a shape, or even comparing two shapes are common in image processing and computer vision. The component labeling processing described earlier in this chapter represents a component as a tree and a set of components as a forest of trees. Although this representation provides excellent support for component labeling it is an inadequate representation for the operations now in view and hence we seek compact representations to efficiently support various component processing algorithms. This section describes the most common techniques for representing a component.

11.5.1   Mask

The most natural representation of a shape is a two-dimensional array M where $M[x][y] == 1$ if location (x, y) is an element of the component and $M[x][y] == 0$ otherwise. In addition, the dimensions of M correspond to the bounding box of the component and the location of the component requires an additional two integer values for the (x, y) coordinate of the upper-left sample. Since each element of the array is either 0 or 1 the array M is implemented as an array of bits and hence this representation is often referred to as a bitmap. The term "bitmask" or simply "mask" is also often used since the mask flags those pixels of a source image that belong to the component.

11.5.2 Run Length Encoding

A component may also be represented through run length encoding. Under run length encoding a component is given by a list of runs where each run represents a sequence of foreground pixels that occur in the same row. Each run is given by three integers that represent the starting location of the run (x, y) in addition to the length of the run. An alternative is to represent a run by three numbers that give the starting location (x, y) and the ending column rather than the length of the run. Consider an array of lists R such that $R[i]$ is a list of runs that occur only in the ith row of the component. Each run could then be given as a pair of values that indicate the first and last columns of the run since the row information is shared by all elements in the list.

11.5.3 Chain Codes

Orienteering is an ancient sport where markers are placed at various locations spread across a large region of countryside. Hikers, or orienteers, must then travel from marker to marker by using only a magnetic compass as a guide and eventually returning to the staring point. Chain coding is an alternative technique for representing a component, which is directly related to orienteering. A chain code defines the contour of a component as each boundary pixel is traversed by following directions aligned with the points of a compass. Chain coding was discovered by Herbert Freeman and is often referred to as Freeman coding. Chain coding gives an attractively compact lossless representation that provides excellent support for many component processing operations. Under chain coding a component is defined by the location of a foreground pixel laying on the boundary of the component. We will refer to this pixel as the starting point P_s. A list of directions is also given that shows how to traverse each of the boundary pixels of the component beginning and ending with P_s.

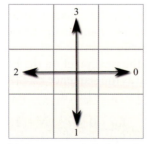
(a) 4-connected chain code.

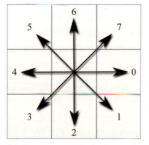
(b) 8-connected chain code.

Figure 11.12. Chain code directions.

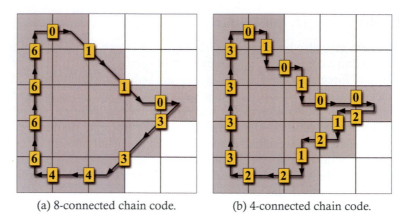

(a) 8-connected chain code.   (b) 4-connected chain code.

Figure 11.13. Chain code examples.

Under 4-connectedness, there are only four possible directions that can be traversed from a boundary pixel p. Those four locations are conventionally given as values 0 through 3, which correspond to east, south, west and north, as shown in Figure 11.12. Under 8-connectedness, there are only eight possible directions that can be traversed and these are also shown in Figure 11.12.

Figure 11.13 gives an example of how chain coding could be used to represent a component. In both (a) and (b) the starting point P_s is given as $(0, 0)$. The 4-connected chain code representation is then given by $\{0101002121223333\}$ while the 8-connected code is given by $\{011033446666\}$. Observant readers may question how chain codes can be used to represent a component having interior holes since the outer contour is the same whether or not the interior hole exists. While beyond the scope of this text, interior holes are typically represented by chain codes that traverse the interior boundary in a counterclockwise direction while exterior contours are traced in a clockwise movement.

The chain code of Figure 11.13 uses an absolute chain code, which is so named since each direction is defined with respect to a global reference point. A differential chain code is a variant on the absolute change code and stores the change in direction rather than the storing the change in position of each pixel traversed. Differential chain codes essentially describe how much an orienteer should spin prior to moving forward into the next pixel on the boundary. If we define the 8-connected absolute chain code as the sequence $c = \{c_0, c_1, \ldots, c_N\}$, the 8-connected differential chain is given by $c_d = \{d_0, d_1, \ldots, d_n\}$ where

$$d_i = \begin{cases} (c_{i+1} - c_i) \mod 8 & \text{for} \quad 0 \leq i \leq N - 1, \\ (c_0 - c_{N-1}) \mod 8 & \text{for} \quad i = N - 1. \end{cases} \quad (11.1)$$

The differential encoding of the component shown in Figure 11.13 is then given by $\{107301020002\}$ from starting point $(0, 0)$. Each entry in the chain gives a multiplier of $45°$ increments of clockwise rotation from the previous direction prior to moving into the next pixel.

One interesting property of differential chain codes is that they are invariant under rotations of $90°$. An absolute chain code that is rotated by $90°$ for example, will produce a vastly different chain code for the rotated component. Rotation of a differential chain code by $90°$ however, will produce the same chain code assuming that the initial direction of travel is also explicitly rotated. Chain codes are not invariant to all rotations since the component boundary will change as a result of resampling. Chain codes are also sensitive to scale since scaling a component will lengthen or shorten the component boundary.

11.6 Component Features

While we have thus far defined a component as a set of connected foreground pixels, it is unlikely that a person in conversation would describe the shapes of an image by listing each of the points in each shape. The more natural approach would be to identify the shape and describe its general location and orientation in the image. You might note, for example, the tall and thin rectangle at the upper-left of the image or the flattened oval at the lower-left of the image. When describing shapes in this manner a large amount of information can be conveyed by a small set of properties that characterize the shape.

Component classification seeks to identify a small set of features that convey essential information about a component in a concise fashion. This set of characteristic properties is known as a feature vector where the term feature is used to denote a property that can be derived from the set of connected pixels that comprise a component. Since the feature vector will contain relatively few values it provides a computationally effective characterization of a component shape. This section describes those features that are commonly used as part of a feature vector.

11.6.1 Geometric Features

The area of a component is given by the number of pixels in the component. If the component is represented as a set of locations, the area is given by the cardinality of the set.

The perimeter of a component is defined as the length of the outer contour. The perimeter is easily derived from the chain code representation by adding the length of each step in the chain. Each of the cardinal directions in a chain code incurs a movement of length 1 while each of the diagonal directions incurs a movement of length $\sqrt{2}$. Equation (11.2) gives a more formal definition where the component is represented by the absolute chain code $c = \{c_0, c_1, \ldots, c_N\}$.

Empirical studies show that Equation (11.2) tends to overestimate the perimeter of the non-discretized component and so a scaling factor of .95 is often applied to produce a more accurate measure of perimeter:

$$\text{Perimeter} = \sum_{i=0}^{N-1} \quad \text{length}(c_i).$$

$$\text{length}(c_i) = \left\{ \begin{array}{ll} 1 & \text{for} \quad c_i = 0, 2, 4, 6, \\ \sqrt{2} & \text{for} \quad c_i = 1, 3, 5, 7. \end{array} \right. \tag{11.2}$$

Note that the area and perimeter of a component are generally invariant to translation and rotation[1] but are obviously sensitive to changes of scale. A larger component will, for example, have a larger perimeter and area than an identically shaped but smaller component. We will now describe two interesting geometric properties that are invariant to translation, rotation, and scale.

Compactness is defined as the relationship between the area of a component to the perimeter. Since we are concerned with developing a scale invariant feature we note that the perimeter of a component increases linearly with an increase of scale while the area increases quadratically. For this reason, the compactness is given as the ratio of the area to the square of the perimeter, as shown in Equation (11.3):

$$\text{Compactness} = \frac{\text{Area}}{\text{Perimeter}^2}. \tag{11.3}$$

Compactness is closely related to the circularity of a shape. The compactness of a perfect circle of any scale is given by $\frac{1}{4\pi}$ and hence when compactness is normalized by this factor we obtain a measure of how similar the component is to a circle. The circularity of a component has a maximum value of 1, which is only achieved by a perfect circle while the minimum circularity approaches zero. A component having low circularity will typically have long thin tendrils that spread across a fairly large region of space:

$$\text{Circularity} = 4\pi \frac{\text{Area}}{\text{Perimeter}^2}. \tag{11.4}$$

11.6.2   Statistical Features

In set theory, the domain or set of all integers is denoted as $\mathbb{Z}$ and hence we say that a coordinate is an element of $\mathbb{Z}^2$ or in other words, a coordinate is an ordered pair of integers. A component can then be defined as the set of coordinates of the pixels that comprise the component. In this presentation we use a characteristic function to define the set of coordinates. The characteristic function accepts a

---

[1]Rotation may introduce relatively small discretization errors.

coordinate and returns 1 if the coordinate is an element of the component and returns a 0 otherwise. Equation (11.5) defines the characteristic function:

$$f(x,y) = \begin{cases} 1 & \text{if } (x,y) \text{ is an element of the component,} \\ 0 & \text{otherwise.} \end{cases} \tag{11.5}$$

Statistical moments are used in many image processing applications ranging from optical character recognition to medical image analysis. These moments describe the point distribution with respect to the horizontal and vertical image axes and provide rich information about the size, shape, and location of the component. In general form, the moments are relative to the image origin and are said to be of some *order*. The general or ordinary moments of a component are then given by Equation (11.6), where p represents the horizontal order of moment μ and q represents the vertical order. Also note that the summation is over all ordered pairs of integers. The equations that follow drop this explicit notation but maintain the range $(x,y) \in \mathbb{Z}^2$ implicitly:

$$m_{pq} = \sum_{(x,y) \in \mathbb{Z}^2} x^p y^q f(x,y). \tag{11.6}$$

The area of a component is given by the moment of order zero, m_{00}, which is also known as the 0th moment. The area of a component is simply the number of pixels that comprise the component as given by Equation (11.7):

$$m_{00} = \sum_{(x,y)} x^0 y^0 f(x,y) = \sum_{(x,y)} f(x,y). \tag{11.7}$$

The centroid, or center of mass, of a component represents the location of the component and is known as the 1st moment. Consider taking a strong, thin, and rigid sheet of metal and trimming it into the shape of some component. The centroid of the component represents the point at which the cutout would be perfectly balanced if it were to be placed atop a sharp pencil point. Note, however, that the centroid of a component may not be contained within the component boundaries as may occur if the component has holes or is not convex. The centroid is given by $(\bar{x}, \bar{y})$ where $\bar{x}$ is the average x coordinate of all points on the component and $\bar{y}$ is the average y coordinate of all points on the component. The centroid is defined in Equation (11.8):

$$m_{10} = \bar{x} = \frac{\sum_{(x,y)} x f(x,y)}{m_{00}}, \quad m_{01} = \bar{y} = \frac{\sum_{(x,y)} y f(x,y)}{m_{00}}. \tag{11.8}$$

While the ordinary moments are dependent on both the location and size of the component the normalized central moments are not dependent upon either the location or the scale of the component. To compute the position independent, or translation invariant, moments we translate the component to the origin and then apply Equation (11.6). Translating a component by $(-\overline{x}, -\overline{y})$ moves the component to the origin, which gives the central movements of Equation (11.9), where μ_{pq} represents the central moment of order (p, q):

$$\mu_{pq} = \sum_{(x,y)} (x - \overline{x})^p (y - \overline{y})^q f(x, y). \tag{11.9}$$

The central moments are also dependent on the size of the component since each movement is a function of the location of each coordinate in the component. In order to normalize for the scale of a component we must introduce a scaling factor into Equation (11.10). Scaling a component by a uniform factor of s in size will scale the central moments by a factor of $s^{(p+q+2)}$. This leads to Equation (11.10), which gives the scale invariant normalized moments of a component:

$$\overline{\mu}_{pq} = \mu_{pq} \left(\frac{1}{\mu_{pq}} \right)^{(p+q+2)/2} \qquad \text{for } p + q \geq 2. \tag{11.10}$$

The orientation of a component is given by the axis that passes through the centroid and is aligned with the longest portion of the component. In much the same way that the centroid of a component can be understood using the metaphor of a thin metal cutout, the orientation can be understood using an analogy drawn from physics. Consider taking the thin cutout and twirling it around some axis lying in the X-Y plane of the component and passing through the centroid. The orientation is the axis that minimizes the inertial momentum that results from spinning the metal sheet in this manner. Consider twirling a baton around an axis passing from one end to the other. The resulting inertial energy is smaller than if the same baton were twirled about an axis perpendicular to that central axis. The orientation can be computed from the moments, as shown in Equation (11.11); note that the resulting angle lies in the range $[-pi/2, pi/2]$ and is only valid when $\mu_{20} \neq \mu_{02}$:

$$\Theta = \frac{1}{2} \tan^{-1} \left(\frac{2\mu_{11}}{\mu_{20} - \mu_{02}} \right). \tag{11.11}$$

The eccentricity of a component is given by the ratio of its longest axis to its shortest axis. Eccentricity is a measure of the elongatedness or thinness of a component. Consider the smallest rectangle that encloses all points in the component and that is aligned with the components orientation. The ratio of the longest side to the smallest side gives the eccentricity. Equation (11.12) gives a definition of eccentricity that is easily computable from the moments and that is closely related to the maximum aspect ratio of the smallest bounding box of a component.

$$\text{Eccentricity} = \frac{\mu_{20} + \mu_{02} + \sqrt{(\mu_{20} - \mu_{02})^2 + 4\mu_{11}^2}}{\mu_{20} + \mu_{02} - \sqrt{(\mu_{20} - \mu_{02})^2 + 4\mu_{11}^2}}. \tag{11.12}$$

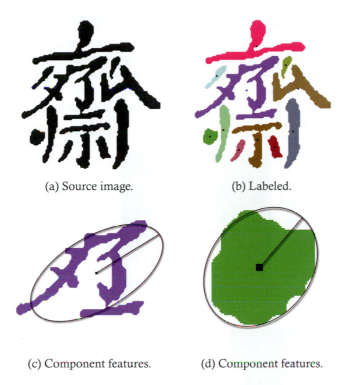

(a) Source image.

(b) Labeled.

(c) Component features.

(d) Component features.

Figure 11.14. Component labeling and statistical features.

The binary source image of Figure 11.14 part (a) shows the signature of the well-known Japanese artist Katsushika Hokusai. The image is labeled in part (b) and the centroid of each component is also indicated by the black dot. Two of the components are highlighted in parts (c) and (d) where the orientation of each component is given by the line extending outward from the centroid and the eccentricity is indicated by the aspect ratio of the oval overlay. The circularity of the component shown in part (c) is .077, which indicates that the component is very unlike a circle. The compactness of the component shown in part (d) is .898, which conveys that the component is circle-like.

11.7 Image Segmentation

Image segmentation refers to partitioning of a digital image into multiple regions where each region is a component. While a binary image is already segmented, a grayscale or color image is not since there is no well-defined background or foreground color. This section gives a very brief overview of the problem of segmenting color and grayscale images.

Consider the definition of connectedness given in Section 11.1, where two pixels p and q are said to be 4-connected if they are 4-adjacent and the same color. While this notion of component is sufficient for binary images since there are only two color classes, background and foreground, it is not adequate for grayscale or color images. For any grayscale or color image the pixels that comprise a single object will likely be similar in color to some degree but certainly not identical in color.

Figure 11.15. Segmentation of color image.

In order to illustrate the idea that components of a grayscale or color image are not identically colored, consider the image of Figure 11.15. In the most general sense, the image is composed of a background component, the dark region behind the subject, and a foreground component, the subject of the portrait itself. In a more specific sense, the subject is composed of a staff, hat, and figure. The subject of the portrait can be further decomposed into a face, torso and right arm as shown in the figure. It is clear that the individual pixels that comprise the hat, for example, are not precisely the same color and hence a more flexible definition of connectedness is required for properly segmenting grayscale and color images.

Segmentation of a color image can be naively implemented by a slight modification of the definition of connectedness. We might relax the requirement that two adjacent pixels be identical in color to require only that two adjacent pixels be *similar* in color. Similarity can be specified, for example by placing a limit on the L_2 distance between two colors such that colors C_1 and C_2 are similar if the L_2 distance between them is less than or equal to some threshold. We might define the hat of Figure 11.15 as the set of pixels that are 4-connected to a user-selected pixel p that has been arbitrarily selected as a representative sample within the hat.

Image segmentation is an active area of research within the image processing community and many different approaches have been developed. Among the most popular are clustering methods such as K-means or region growing methods in addition to newer techniques based on neural networks. These approaches are beyond the scope of this text and are not discussed further.

11.8   Exercises

1. Segment the following binary image using N_4, N_d, and N_8 connectedness. Each component of a segmentation should be given as a list of pixel coordinates that belong to the component.

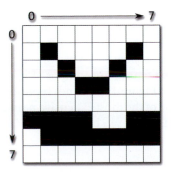

2. Using the N_4 segmentation of 1, give the centroid, perimeter, compactness, and circularity of the two largest components.

3. Using the N_8 segmentation of 1, give the centroid, perimeter, compactness, and circularity of the two largest components.

4. Draw the results of both eroding and dilating the image of problem 1 with a 3×3 square mask.

5. Draw the results of both eroding and dilating the image of problem 1 with a 3×3 plus-shaped mask.

6. Give the chain code representation for each component in the N_4 and N_8 segmentation of problem 1.

7. Locate an image of handwritten text and load this image into Pixel Jelly. Use any part of the application to generate a segmentation of the image so that each letter is a connected component. Note that it may not be possible to accomplish this task completely since some letters will be so connected to others as to be impossible to automatically separate.

Artwork

Figure 11.4, 11.8, and 11.11. *Mona Lisa* by Leonardo da Vinci (1452–1519). See page 170 for a brief biography of the artist. Da Vinci's *Mona Lisa*, from which the image of Figure 11.4 and others is derived, is unquestionably the most famous and most copied portrait painting of all time. *Mona Lisa* (also known as *La Gioconda*) is a 16th-century portrait, which at the time of this writing was owned by the French government and hangs in the Musée du Louvre in Paris, France, with the title *Portrait of Lisa Gherardini, Wife of Francesco del Giocondo*.

Figure 11.15. *Portrait of a Halberdier* by Jacopo Pontormo (1494–1557). Pontormo was an Italian painter specializing in portraiture following the mannerist school. Pontormo was apparently orphaned at a young age but had the great fortune to apprentice under the likes of Leonardo da Vinci, Mariotto Albertinelli, and other great artists of his time. Much of his art focused on religious themes as illustrated by *The Deposition from the Cross*, considered to be one of his best works, where brilliantly colored figures are composed in a complex and ethereal form. While the illustration of Figure 11.15 is not representative of his style as a whole, it is notable as being one of the most expensive paintings in history. The *Portrait of a Halberdier* was purchased in 1989 by the Getty Museum for the sum of US$35.2 million.

Advanced Programming 12

12.1 Overview

This section presents issues that are more related to implementation issues of image processing than to the concepts behind image processing itself. In this chapter we briefly describe how concurrency can be used to provide significant speedups, and also give an introduction to the JAI library, a standard extension to Java that supports the development of high performance image processing applications.

12.2 Concurrency

The execution speed of many of the image filters presented throughout this text can be significantly improved through the use of concurrency. Concurrency, or parallelization, occurs when more than one process within a computing system is executing at the same time. While concurrency is possible on simple single-cpu, single-core computing platforms, significant software speedup can only be realized on computing systems that have multiple processors or multiple cores on a single processor.

Java provides elegant support for concurrency through both language-level features and a well designed set of classes. We will first describe basic techniques for generating concurrent threads of execution in Java and then demonstrate how concurrency can be used to increase the performance of a certain class of image processing operations.

A Java *process* is a heavyweight object that controls a complete set of system-level resources including its own memory space. A process can be understood as a single Java program. A *thread* executes within the context of some process and shares the resources and memory space of that process. The creation and control of Threads is the primary mechanism for supporting concurrency in Java. The

Thread
+void run()
+void join()
+void start()

Figure 12.1. Thread methods (partial listing).

Thread class contains methods for creating and controlling a thread of execution where the most important methods for our purposes are shown in the UML class diagram of Figure 12.1.

When a Thread is created, it does not begin execution but is simply prepared for execution. The Thread constructor simply initializes the data structures necessary to establish an execution environment without actually allocating the system resources necessary for a process to execute. Not until the start method is invoked is the execution environment fully initialized by allocating system resources after which the run method is executed within the environment. The run method is the code that the thread executes in parallel with other active threads. The default run method is empty and hence this class is designed to be subclassed where the run method is overridden in some desired fashion. If we assume that t is a Thread then t.start() returns immediately to the caller but initiates the execution of t.run(), which may itself not be completed until a much later time.

The join method allows one thread to wait for the completion of another. If we assume that t is a Thread then t.join() causes the currently executing thread to stop and wait for t to complete. When t has finished execution then the currently executing thread awakens and continues execution.

In image processing it is common to implement concurrent operations by splitting an image into rectangular tiles that are processed simultaneously by various threads of execution. As a first step towards the design of a flexible and general framework for concurrent image processing operations we will design a thread that iterates over some subset of the *tiles* of an image and applies a BufferedImageOp to each of the tiles it is assigned.

We first develop the class that is responsible for tiling an image and producing the tiles in sequence. The ImageTiler class shown in Listing 12.1 is a complete listing of a thread-safe iterator producing a sequence of tiles covering some source image.

The constructor accepts a BufferedImage and the size of a single tile as given by the width and height parameters. Tiles are generated as Rectangles in this implementation where the tiles are made available in raster-scan order through the next method. The next method returns a null value in the event that no more tiles are available in the scan otherwise the next tile in the scan is produced.

```java
public class ImageTiler implements Iterator<Rectangle> {
  private int tileWidth,  tileHeight, tilesPerRow, tileRows;
  private int currentTile;

  public ImageTiler(BufferedImage src, int w, int h) {
    tileWidth = w;
    tileHeight = h;
    int srcWidth = src.getWidth();
    int srcHeight = src.getHeight();
    currentTile = -1;
    tilesPerRow = (int) Math.ceil(srcWidth / (double) tileWidth);
    tileRows = (int) Math.ceil(srcHeight / (double) tileHeight);
  }

  public synchronized boolean hasNext() {
    return currentTile < tilesPerRow * tileRows - 1;
  }

  public synchronized Rectangle next() {
    Rectangle result = null;
    if(hasNext()) {
      currentTile++;
      result = new Rectangle(getX(), getY(), tileWidth, tileHeight);
    }
    return result;
  }

  private int getX() {
    return (currentTile % tilesPerRow) * tileWidth;
  }

  private int getY() {
    return (currentTile / tilesPerRow) * tileHeight;
  }

  public synchronized void remove() {
    throw new RuntimeException();
  }
}
```

Listing 12.1. ImageTiler.

The key feature of the ImageTiler is that of thread safety. Each of the public methods is synchronized, which ensures that if multiple threads attempt to simultaneously access one of these methods on the same ImageTiler object, only one thread at a time is granted permission to proceed. This control is necessary to ensure that the internal state of the ImageTiler is not mangled by allowing multiple threads to step through the body of the next method in an unpredictable order. A synchronized method can be viewed as an atomic action, an entire series of instructions that effectively occurs as a single instruction.

Listing 12.2 defines a specialized thread that uses an `ImageTiler` to obtain tiles from a source image. When a `OpThread` is created the constructor requires a non-null `BufferedImageOp`, source and destination images, and an `ImageTiler`. When the thread is executed, the run method is invoked such that the tiles generated by the `ImageTiler` are used to create sub-images of both the source and destination. These sub-images are then processed in the conventional manner through use of the `BufferedImageOp` filter method.

```java
1  public class OpThread extends Thread {
2    private BufferedImage src, dest;
3    private BufferedImageOp op;
4    private ImageTiler scanner;
5
6    public OpThread(BufferedImageOp op,
7                    BufferedImage src,
8                    ImageTiler scanner,
9                    BufferedImage dest) {
10     this.op = op;
11     this.src = src;
12     this.scanner = scanner;
13     this.dest = dest;
14   }
15
16   public void run() {
17     Rectangle srcBounds = src.getRaster().getBounds();
18     Rectangle tile = null;
19     while ((tile = scanner.next()) != null){
20       tile = tile.intersection(srcBounds);
21       BufferedImage srcTile =
22         src.getSubimage(tile.x, tile.y, tile.width, tile.height);
23       BufferedImage destTile =
24         dest.getSubimage(tile.x, tile.y, tile.width, tile.height);
25       op.filter(srcTile, destTile);
26     }
27   }
28 }
```

Listing 12.2. `OpThread`.

Proper understanding of the `getSubimage` method is crucial. While the `getSubimage` method creates a new `BufferedImage` representing a portion of some parent image, the `Raster` of the sub-image and the `Raster` of the parent image are the same. This implies that changing any sample of the sub-image also changes the corresponding sample of the parent image. Hence, each of the sub-images generated in the run method corresponds to portions of either the source or destination images.

We now take one further step by writing code that creates multiple OpThreads and sets them up to iterate over the same source using the same tiler to pour

```java
public class ConcurrentOp extends NullOp {
    private BufferedImageOp op;
    private int threadCount, tileWidth, tileHeight;

    public ConcurrentOp(BufferedImageOp op,
                        int tileWidth,
                        int tileHeight,
                        int threadCount) {
        this.op = op;
        this.threadCount = threadCount;
        this.tileWidth = tileWidth;
        this.tileHeight = tileHeight;
    }

    public ConcurrentOp(BufferedImageOp op, int tileWid, int tileHt) {
        this(op, tileWid, tileHt, Runtime.getRuntime().availableProcessors());
    }

    @Override
    public BufferedImage filter(BufferedImage src, BufferedImage dest) {
        if(dest == null) {
            dest = op.createCompatibleDestImage(src, src.getColorModel());
        }

        ImageTiler scanner = new ImageTiler(src, tileWidth, tileHeight);
        Thread[] threads = new Thread[threadCount];
        for(int i=0; i<threadCount; i++){
            threads[i] = new OpThread(op, src, scanner, dest);
            threads[i].start();
        }

        for(Thread t : threads) {
            try {
                t.join();
            } catch (InterruptedException ex) {
            }
        }

        return dest;
    }
}
```

Listing 12.3. ConcurrentOp.

output into the same destination image. In addition, this code is embodied using the conventional `BufferedImageOp` paradigm. The complete class is shown in Listing 12.3.

A `ConcurrentOp` is essentially a thin wrapper that takes any non-concurrent `BufferedImageOp` and automatically parallelizes the filter method. As shown in listing 12.3, the constructor accepts a non-concurrent operation, the desired tile sizes, and the desired number of threads to use when filtering. The second constructor, shown on line 15, assumes that the number of desired threads is equivalent to the number of processors available on the system. The number of available processors can be found dynamically in Java through the use of the `Runtime` class as illustrated in line 16.

The `filter` method begins in normal fashion by constructing a destination image if one is not provided. The code continues by constructing a single `ImageTiler` that will provide tiles of the source upon request. An array of threads is then constructed and initialized in line 28 such that each `OpThread` object uses the same `BufferedImageOp` to filter the same source image using the same `ImageTiler` to produce the same destination image. After construction, each thread is then started as shown on line 29 of Listing 12.3.

In a sense, the filter method is complete at that point. Each of the created threads will execute until no more tiles of the source image remain at which time the destination image will contain the correct result. Since, however, the filter method requires that the destination image must be returned, the filter method must wait until all threads have completed prior to returning the destination image. This is done by joining each of the created threads, that is, by waiting for each of the threads to complete, as shown on line 34.

The code in Listing 12.3 assumes that the supplied `BufferedImageOp` can be safely parallelized over tiles of an image. The assumption is that there is no intrinsic reliance on the order in which pixels are processed or on the location of the image pixels or the overall dimensionality of the source or destination image. While most of the image processing operations presented in this text are parallelizable in this way, some filters rely on the order of pixel scanning or on image geometry and hence produce faulty results when parallelized using this technique.

Figure 12.2 gives an illustration of the improper use of the `ConcurrentOp` class. The source image of part (a) is filtered with a `KaleidoscopeOp` in the standard non-concurrent method as shown part (b). When this filter is applied in the concurrent fashion developed in the `ConcurrentOp` class the result is obviously incorrect, as shown in part (c). It is important to note that while certain operations may not be parallelizable using the technique described in this text, it is likely that specialized code can be developed to parallelize most image filters. We have here described a technique that is relatively flexible and which can be properly applied to a reasonably large set of image processing filters.

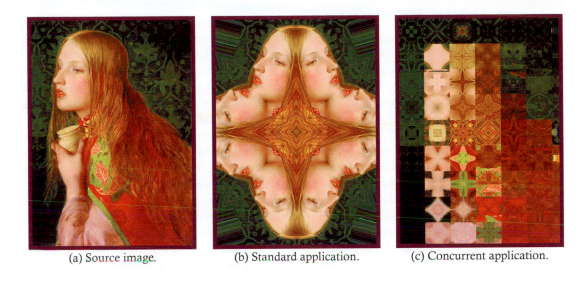

(a) Source image.   (b) Standard application.   (c) Concurrent application.

Figure 12.2. Improper use of concurrency.

12.3   JAI

Java provides support for the development of media-rich applications through the Java Media suite. The Java Media suite is a collection of related software packages that provide support for various multimedia components. The Java 2D package supports two-dimensional graphics and basic image processing features. The Java 3D package provides support for the creation of three-dimensional graphics while the Java Media Framework supports components that control time-based data such as audio and video. The Java Advanced Imaging (JAI) package is part of the Java Media suite and provides support for sophisticated image processing development through an extremely flexible and optimized set of classes. While JAI is not part of the standard Java distribution, it is a freely available standard extension and is as good or better than many commercially available C and C++ image processing libraries.

The JAI provides a much richer set of features than the standard Java 2D classes described throughout the earlier portions of this text. Among the most noteworthy features are full support for floating point rasters, a larger range of image decoders and encoders, a larger range of color models, multiple-source filters, deferred execution, and support for distributed and remote imaging.

When image processing filters are executed within the JAI framework the resulting code can run in either accelerated or Java mode. The JAI package, when it is installed on a particular platform, may include a set of libraries written in C and is highly optimized for the platform on which the library is installed. These libraries take advantage of hardware acceleration for many of the built-in processing functions and therefore provide significant speedup for image processing applications. Whenever a JAI filter is executed, the JAI package first determines whether these optimized native libraries have been installed. If the libraries are available, the filter is run in accelerated (or native) mode, meaning that the code in those libraries is utilized to perform the filter. If the libraries are not available, the filter will execute properly but in non-accelerated or Java mode.

While the JAI library provides elegant support for the development of high performance image processing, it is correspondingly complex in its design and hence largely beyond the scope of this text. Nonetheless, we will outline in this section the way in which JAI applications are modeled and implemented.

In the JAI suite, an image processing filter is viewed as a chain of individual operations. This chain is modeled as a directed acyclic graph (DAG) where each node corresponds to an operation and each arrow relates the flow of data between operations. While each node in the graph is an operation, it is interesting to note that each node is also *the image that results from applying the operation*. Consider Figure 12.3, which illustrates how a typical JAI filter is constructed. In this example the node labeled as "Image 1" represents an operation that loads an image from a file. The operation itself is given in the dashed box within the node and the node itself represents the image that results from the operation. This image is then fed into the node labeled "Image 2" as the source image. The Image 2 node represents a minimum rank filter using a 3×3 square mask and the generated image is then fed into the node labeled Image 3 as one of the two sources to that node. The Image 3 node represents the addition of the first two images.

The central classes of the JAI library are RenderedOp, ParameterBlockJAI, and JAI. The RenderedOp class represents a node in a chain of operations. As the name implies, a RenderedOp is an image processing operation that has been executed to produce a rendered image. A RenderedOp maintains the name of the operation it represents, a ParamaterBlock that serves as a collection of parameters controlling the operation and a vector of source images on which the

Figure 12.3. Typical JAI filter representation.

operation should be performed. The JAI class is a convenience class following the factory pattern whose purpose is to simplify the construction of JAI chains through a set of static create methods.

Listing 12.4 shows a code fragment that constructs the JAI chain illustrated in Figure 12.3. Each call to JAI.create produces a RenderedOp object representing an operation where the operation is identified by the name used in the parameter block constructor and is controlled by the values within the parameter block. A ParameterBlock is a sequential list of the parameters required by an operation. Each of the individual parameters within the block are named and hence the parameters can be defined by the setParameter method, which takes the name of a parameter followed by its actual value.

Consider the *FileLoad* operation on line 3. There is a single parameter that controls this operation and that parameter has the name *filename*. This parameter is established by the setParameter method of line 4. ParameterBlocks are designed to generalize the construction of image filters. While some image processing operations don't require any parameter, (inversion, for example) image processing operations in general do require controlling parameters. In Java 2D, a BufferedImageOp class represents an image filter where the controlling parameters are provided to the constructor. This requires construction of specialized BufferedImageOp subclasses with specialized constructors for each. In JAI, this process is unified through the use of ParameterBlocks such that all image processing operations have a set of parameters that are stored within the block and accessed within the code body where filtering is actually performed. The edges

```
1   RenderedOp op0, op1, op2;
2
3   ParameterBlockJAI parameters0 = new ParameterBlockJAI("FileLoad");
4   parameters0.setParameter("filename","d:/Flower3.jpg");
5   op0=JAI.create("FileLoad", parameters0, null);
6
7   ParameterBlockJAI parameters1= new ParameterBlockJAI("MinFilter");
8   parameters1.setParameter("maskShape", MinFilterDescriptor.MIN_MASK_SQUARE);
9   parameters1.setParameter("maskSize",3);
10  parameters1.addSource(op0);
11  op1=JAI.create("MinFilter", parameters1, null);
12
13  ParameterBlockJAI parameters2= new ParameterBlockJAI("Add");
14  parameters2.addSource(op0);
15  parameters2.addSource(op1);
16  op2=JAI.create("Add", parameters2, null);
```

Listing 12.4. Creating a JAI chain.

of the graph are established through the addSource methods. Note that the *min filter* operation has a single source that is given as op0 while the *add* filter has two sources that are specified in lines 14 and 15.

Artwork

Figure 12.2. *Mariya Magdalena* by Anthony Frederick Augustus Sandys (1829–1904). Frederick Sandys was a British painter, illustrator, and caricaturist. His largest body of work is found in the illustrations he penned for periodicals such as *Good Words* and *Once a Week* among others. While he was not a prolific painter he was a gifted painter who focused on mythic and Biblical themes. The illustration of Figure 12.2 is the artists depiction of Mary of Magdala, or Mary Magdalene. Mary Magdalene is described in the Biblical texts as a woman from whom Jesus himself cast out seven demons (Luke 8:2). She subsequently became a devoted disciple of Jesus who witnessed his crucifixion (Mark 15:40) and was the first to see him after his resurrection (Mark 16:9). She has unfortunately been misidentified as the prostitute who washed the feet of Jesus as recorded in Luke 7:36–50. Artists renditions tend to adopt this erroneous view of her identity and thus present her with long flowing red hair or holding a jar of perfume, as Sandys does in Figure 12.2.

Floating Point Rasters

When constructing a `BufferedImage` a client must choose the color model and transfer type for the underlying data. While Java provides full support for integer types it unfortunately provides little support for the floating point types. A `BufferedImage` that uses a floating point data buffer will not be properly rendered when displayed within a `Swing` component for example. Nonetheless, programmers are able to construct images that use floating point buffers and process them with operations of their own construction.

The `ConvolutionOp` presented in this text is an example of how to construct a floating point data buffer. Whenever an image is convolved with some kernel, the result is always a floating point image which requires further post-processing in order to properly display. The `ConvolutionOp` overrides the `createCompatibleDestImage` method to construct the required destination type. This method is shown in Listing A.1.

```java
public BufferedImage createCompatibleDestImage(BufferedImage src,
                                               ColorModel model) {
    int totalSize = src.getWidth() * src.getHeight();
    int[] offsets;

    if(src.getRaster().getNumBands() == 3) {
        offsets = new int[]{0, totalSize, totalSize*2};
    } else {
        offsets = new int[]{0};
    }

    SampleModel sampleModel =
        new ComponentSampleModel(DataBuffer.TYPE_FLOAT,
                                 src.getWidth(),
                                 src.getHeight(),
                                 1,
                                 src.getWidth(),
                                 offsets);
```

```
19
20    int numSamples = src.getWidth() *
21                        src.getHeight() *
22                        src.getRaster().getNumBands();
23    DataBuffer dataBuffer = new DataBufferFloat(numSamples);
24    WritableRaster wr = Raster.createWritableRaster(sampleModel,
25                                                      dataBuffer,
26                                                      null);
27    ColorModel colorModel =
28      new ComponentColorModel(model.getColorSpace(),
29                                false,
30                                true,
31                                Transparency.OPAQUE,
32                                DataBuffer.TYPE_FLOAT);
33
34    return new BufferedImage(colorModel, wr, true, null);
35  }
```

Listing A.I. Creating a floating point image.

Construction of the SampleModel on line 12 requires that the type of the data buffer be given, in this case a TYPE_FLOAT along with the dimensions and offsets. In this example, the offsets indicate the indices within the buffer that correspond to the bands of the buffer.

Scanners

B

The basic `for` loop was augmented in Java 1.5 to support convenient iteration over collections. We will denote this newer type of `for` loop as a `for-each` loop within this discussion.

Consider writing a loop to iterate over the elements of a collection and processing the elements as they are visited. Assume that a collection of some type has been constructed to which data has been inserted. In this case, Listing B.1 shows how the old-style `for` loop would be used to iterate over the collection.

```
1  for(Iterator<TYPE> iter = collectionObject.iterator(); iter.hasNext(); ) {
2    TYPE var = iter.next();
3    // process var
4  }
```

Listing B.1. Old style `for` loop.

The old style `for` loop explicitly makes use of an `Iterator` object that sequentially visits the elements in the collection through successive calls to `next`. While this code pattern is very common and powerful it is made more convenient by the `for-each` loop shown in Listing B.2.

```
1  for(TYPE var : collectionObject) {
2    // process var
3  }
```

Listing B.2. For-each loop.

In the `for-each` loop there is no reference to an iterator since the iterator is implicitly constructed and used to navigate through the collection. In more technical terms, the only requirement on the collection object is that it must implement the `Iterable` interface which is shown in Listing B.3.

```
1  public interface Iterable<T> {
2    public Iterator<T> iterator();
3  }
```

Listing B.3. `Iterable`.

The `Iterable` interface is a contract that all implementing classes will provide an iterator through the `iterator` method. Of course this is necessary since the `for-each` loop will take the collection object, request an iterator via the `iterator` method, and then navigate through the elements using the iterator itself. Knowing this we can now implements image scanners.

First we develop the `ImageScanner`, as shown in Listing B.4. Note that an image scanner is an abstract class that is also an iterator, producing `ColRowBand` objects and that it also implements the `Iterable` interface. Concrete subclasses must provide concrete implementations of the `next` and `hasNext` methods.

```
1  abstract public class ImageScanner
2          implements Iterable<ColRowBand>, Iterator<ColRowBand> {
3    protected static UnsupportedOperationException exception =
4          new UnsupportedOperationException();
5    protected ColRowBand current;
6    protected int numBands;
7    protected Rectangle region;
8
9    public ImageScanner(Rectangle region, int numBands) {
10     this.region = region;
11     this.numBands = numBands;
12     current = new ColRowBand(0,0,0);
13   }
14
15   public ImageScanner(Rectangle region) {
16     this(region, 1);
17   }
18
19   public boolean hasMoreElements() {
20     return hasNext();
21   }
22
23   public ColRowBand nextElement() throws NoSuchElementException {
24     return next();
25   }
```

```
26
27    public void remove() {
28        throw exception;
29    }
30
31    public Iterator<ColRowBand> iterator() {
32        return this;
33    }
34 }
```

Listing B.4. `ImageScanner`.

Listing B.5 gives the complete source code for the `RasterScanner`. The central task of this object is to manage indices that correspond to a raster scan of an image. This class implicitly assumes that the data is rasterized in row-major format such that a single integer value suffices for keeping track of the column, the row, and even the band of the image being scanned. The `next` method is the most important for our purposes as it increments a counter and then partitions the counter into corresponding column, row, and band index.

```
1  public class RasterScanner extends ImageScanner {
2      protected int index;
3      protected int maxIndex;
4
5      public RasterScanner(BufferedImage image,
6                            Rectangle region,
7                            boolean isBanded) {
8          super(region, isBanded ? image.getRaster().getNumBands() : 1);
9          maxIndex = region.width * region.height * numBands;
10         index = -1;
11         current.col = getCol();
12         current.row = getRow();
13         current.band = getBand();
14     }
15
16     public RasterScanner(BufferedImage image, boolean isBanded) {
17         this(image,
18              new Rectangle(0, 0, image.getWidth(), image.getHeight()),
19              isBanded);
20     }
21
22     public int getCol() {
23         return (index / numBands) % region.width;
24     }
25
26     public int getRow() {
27         return index / (numBands * region.width);
28     }
29
```

```
30    public int getBand() {
31      return index % numBands;
32    }
33
34    public ColRowBand next() {
35      if (!hasMoreElements()) {
36        throw new NoSuchElementException();
37      }
38
39      index = index + 1;
40      current.row = getRow();
41      current.col = getCol();
42      current.band = getBand();
43      return current;
44    }
45
46    public boolean hasNext() {
47      return index < maxIndex − 1;
48    }
49 }
```

Listing B.5. RasterScanner.

References

[Ataman et al. 80] E. Ataman, V. Aatre, and K. Wong. "A Fast Method for Real-Time Median Filtering." *Acoustics, Speech and Signal Processing, IEEE Transactions on* 28:4 (1980), 415–421.

[Cooley and Tukey 65] James W. Cooley and John W. Tukey. "An Algorithm for the Machine Calculation of Complex Fourier Series." *Mathematics of Computation* 19:90 (1965), 297–301.

[Gonzalez and Woods 92] Rafael C. Gonzalez and Richard E. Woods. *Digital Image Processing*, Third edition. Reading, MA: Addison-Wesley, 1992.

[Huang et al. 79] T. Huang, G. Yang, and G. Tang. "A Fast Two-Dimensional Median Filtering Algorithm." *Acoustics, Speech and Signal Processing, IEEE Transactions on* 27:1 (1979), 13–18.

[Kandel et al. 00] E. R. Kandel, J. H. Schwartz, and T. M. Jessel. *Principles of Neural Science*, Fourth edition. New York: McGraw-Hill, 2000.

[Oyster 99] C. W. Oyster. *The Human Eye: Structure and Function*. Sunderland, MA: Sinauer Associates, 1999.

[Ridler and Calvard 78] T. W. Ridler and S. Calvard. "Picture Thresholding Using an Iterative Selection Method." *IEEE Transactions on Systems, Man and Cybernetics* 8:8 (1978), 630–632.

[Roorda et al. 01] A. Roorda, A. B. Metha, P. Lennie, and D. R. Williams. "Package Arrangement of the Three Cone Classes in Primate Retina." *Vision Research* 41:10–11 (2001), 1291–1306.

Index